PRAISE FOR *Confronting*

"This book not only explains the essence of the subject but provides a penetrating analysis of the wider political, military, and economic implications. The second half of the Oil Age now dawns and will be marked by the decline of oil production and all that depends upon it, including especially transport, trade, and agriculture. The book ends with a list of sensible new policy proposals with which to face this turning point of historic magnitude."
—**Colin Campbell, Ph.D.**, former oil exploration geologist, Texaco, British Petroleum; exploration manager, Total; former consultant to Shell, Statoil, Mobil, and Amerada; former Executive V.P., Petrofina; author of many books and publications on oil and gas depletion

"Mike Ruppert has an unblemished track record for saying things that are incendiary, outrageous, shocking—and true. Our new president needs desperately to hear the uncomfortable message of this book about energy and the economy, and so do the rest of us."
—**Richard Heinberg, Ph.D.**, author of *The Party's Over*, *Peak Everything*, *The Oil Depletion Protocol* and senior fellow, Post Carbon Institute

"Mike Ruppert has been at the forefront of speaking and writing about the grim reality that the world's crude-oil output is peaking or has already peaked and will soon begin what could be swift declines over the next decade or two. The world needs to pay careful attention to the multiple risks this event will usher in. Thanks to Ruppert's new book, readers around the world will have access to his well-written work."
—**Matthew R. Simmons**, Chairman of Simmons & Company and author of *Twilight in the Desert: The Coming Saudi Oil Shock and the World Economy*

"Ruppert confronts the stark realities of a world of declining oil production, poses vital questions of our complex oil-dependent supply chains, and challenges us—people and politicians alike—to build a sustainable world with what remains of our resources."
—**Julian Darley**, author of *High Noon for Natural Gas*, founder of Post Carbon Institute

"All I can say is, "Yikes!" This is a book everyone should read. Mike Ruppert is my friend. And sometimes I remind him, in a way that only a friend can, that my perspective is colored by my own distinct experiences as an informed woman of color in the United States. And, frankly, that means that some of what is between these covers makes me cringe; but it is exactly this substance, actively suppressed in proposed national and international gatherings, that we human beings must debate and resolve, or else we will find Dr. King's admonition, once again, to be true: "We must learn to live together as brothers or perish together as fools." We know Mike Ruppert because he became a whistleblower and told us some inconvenient truths—about crack cocaine, 9/11/01, and now this: how to step back from the brink of human disaster.

It is clear that Mike and I are headed toward the same destination, despite our differences. This book lands Mike exactly where I am—outside the box of political orthodoxy, but well within the space of policy advocacy that is representative of critical thinking, rational analysis, and authentic leadership. Mike Ruppert dares to go where our elected leaders seem afraid to take us. In the end, however, if we are to salvage our own human dignity, either our "leadership" must catch up with us or we must become and nurture a new generation of leaders."

—**Cynthia McKinney**, six-term member, U.S. House of Representatives, Green Party Presidential Candidate, 2008

"America's most courageous and fearless investigative reporter exposes the root causes of the financial meltdown. Our new president should read this book for his next intelligence briefing."

—**Mark Robinowitz**, author of *Peak Oil Wars and Global Permaculture Solutions*, PeakOilWars.org, GlobalPermaculture.org

"If ever there was a need for a particular book at a particular time, it's this book now."

—**Jenna Orkin**, World Trade Center Environmental Organization

C O N F R O N T I N G

COLLAPSE

The Crisis of Energy and Money in a Post Peak Oil World

A 25-Point Program for Action

MICHAEL C. RUPPERT

Author of *Crossing the Rubicon*

CHELSEA GREEN PUBLISHING
WHITE RIVER JUNCTION, VERMONT

Confronting Collapse was first
published in 2009 as
A Presidential Energy Policy:
Twenty-five Points Addressing the
Siamese Twins of Energy and Money.

Project Manager: Patricia Stone
Indexer: Peggy Holloway
Designer: Peter Holm,
        Sterling Hill Productions

Chelsea Green Publishing is committed to preserving
ancient forests and natural resources. We elected to print
this title on 100-percent postconsumer recycled paper,
processed chlorine-free. As a result, for this printing, we
have saved:

147 Trees (40' tall and 6-8" diameter)
47 Million BTUs of Total Energy
14,000 Pounds of Greenhouse Gases
67,428 Gallons of Wastewater
4,094 Pounds of Solid Waste

Chelsea Green Publishing made this paper choice because
we and our printer, Thomson-Shore, Inc., are members
of the Green Press Initiative, a nonprofit program dedicated to supporting authors, publishers, and suppliers
in their efforts to reduce their use of fiber obtained
from endangered forests. For more information, visit:
www.greenpressinitiative.org.

Environmental impact estimates were made using the Environmental Defense Paper Calculator.
For more information visit: www.papercalculator.org.

Printed in the United States of America
First Chelsea Green printing December, 2009
10 9 8 7 6 5 4 3 2 1   09 10 11 12 13

Our Commitment to Green Publishing

Chelsea Green sees publishing as a tool for cultural change and ecological stewardship. We
strive to align our book manufacturing practices with our editorial mission and to reduce
the impact of our business enterprise in the environment. We print our books and catalogs
on chlorine-free recycled paper, using vegetable-based inks whenever possible. This book
may cost slightly more because we use recycled paper, and we hope you'll agree that it's
worth it. Chelsea Green is a member of the Green Press Initiative (www.greenpressinitia-
tive.org), a nonprofit coalition of publishers, manufacturers, and authors working to protect
the world's endangered forests and conserve natural resources. Confronting Collapse was
printed on Rolland Enviro Natural, a 100-percent postconsumer recycled paper supplied by
Thomson-Shore.

Library of Congress Cataloging-in-Publication Data is available upon request.

Chelsea Green Publishing Company
Post Office Box 428
White River Junction, VT 05001
(802) 295-6300
www.chelseagreen.com

This book is dedicated to the great and courageous visionary Marion King Hubbert and to President James Earl Carter, Jr.

King, you were the prophet.

Jimmy, you led. You told the truth about energy. That record can never be diluted, and it will not be forgotten. You did not let us down, Jimmy. We let ourselves down.

But then again . . . we were conditioned to.

I must not fear.
Fear is the mind-killer.
Fear is the little death that brings total obliteration.
I will face my fear.
I will permit it to pass over me and through me.
And when it has gone past, I will turn the inner
     eye to see its path.
Where the fear has gone there will be nothing.
Only I will remain.

—FRANK HERBERT, *Dune*

TABLE OF CONTENTS

Michael Ruppert does not mince words in this stirring and uncompromising book on the vital issue of global energy and economics. He addresses some simple but widely ignored concepts relating to the critical role of oil and gas in the modern world. First, they are finite resources formed in the geological past and are, therefore, subject to depletion. Second, they have to be found before they can be produced, the peak of which is long past.

Ruppert considers the broader implications of this realization in terms of a finite Oil Age. The Oil Age started only 150 years ago, precipitating the rapid expansion of industrialization, transportation, technology, etc., fuelled by this cheap-source energy flowing from the ground. But now, at the dawn of the second half of this age, we face a decline of production and all that depends upon it. The economic and political consequences of this turning point for mankind are colossal, demanding far-reaching political responses, as the book discusses. Many claims have been made that new technology will counter the natural decline, yet the arguments overlook the irony that the better the technology, the faster the depletion.

Having laid out the basic facts, Ruppert turns to related subjects, including foreign policy and the invasion of Iraq, the hopes for renewable energy substitutes, the impact on farming and population, and the nature of Money. Though the central theme of this book focuses on the impact of the decline of oil on the American economy, it also explores its global repercussions and puts forth twenty-five sensible recommendations by which the U.S. government can react to the unfolding situation.

This perceptive, stimulating, and readable book grapples with a subject of critical importance. It deserves a place on the bookshelves of every conscientious thinker and actor from the school teacher to the chief executive, the bishop to the politician and world leader.

Colin Campbell, Ph.D.

Colin Campbell is a former oil exploration geologist at Texaco and British Petroleum, a former exploration manager at Total, a former consultant to Shell, Statoil, Mobil, and Amerada, a former executive vice president at Petrofina, the author of many books and publications on oil and gas depletion, and the co-founder of The Association for the Study of Peak Oil.

PREFACE

This book was originally titled *A Presidential Energy Policy: Twenty-five Points Addressing the Siamese Twins of Energy and Money*. The first-submission draft of the book was completed in late 2008.

The book that you see here is essentially that same book, finished professionally, with topical updates that do not correct or change the substance or meaning of what I first wrote with such urgency. As the global economic collapse has unfolded, we have seen that I was spot-on in reading a chilling map of humanity's future if we do not start making fundamental changes in the human paradigm right now. The twenty-five points remain unchanged from earlier versions. The only additions are references to news stories from 2009 that establish that what I wrote then was accurate and has evolved from prediction to history.

I do not wish to delve deeply into the gauntlet this book has run from its writing until it reached the shelves in bookstores. In the scheme of things, that story is a sidebar to a sidebar and I would rather not occupy the reader's mind with that struggle here. Suffice it to say that there was a collective mindset that could not bear and did not want to see these increasingly self-evident truths reach the public consciousness. That mindset has failed and is continuing to lose its grip on public discourse as the old paradigm gives way to the new.

Rather, I would like the reader to know that—like blades of grass growing through a sidewalk—inspired truth, backed by commitment and dedication, can find its way into the sunlight. The saga of how this book came to life in its present form gives hope to all those who look at what is called for in its pages and feel inadequate. I believe that books, like songs, are living things, imbued with spirit and the sweat of the author and those who supported him or her. Thanks to the success of Bluemark Films' *CoLLapse*, the information contained herein will reach more people and spark more change than I could ever have hoped to see in my lifetime.

The only big question that remains is how much time we have and how much we can get done while we have the time, the means, and the will to do it. The title *Confronting Collapse* is especially appropriate because my entire life has repeatedly taught me that the path of a true spiritual warrior is to run toward, and not away from, the things of which we are most afraid.

ACKNOWLEDGMENTS

First and foremost, let me both acknowledge and thank all of the great Peak Oil/Sustainability activists and friends who have labored so long and hard to awaken our collective consciousness. It was your dedication, research, insight, sacrifice, critical contributions to this work, and your incredible care that inspired me to keep writing, no matter what. You'll find much of yourselves in these pages, having sat beside and inside me as I wrote each word. It was you who taught me.

To Dale Allen Pfeiffer, Richard Heinberg, Colin Campbell, Matt Simmons, Professor Al Bartlett, Professor David Pimentel, Robert Hirsch, Ali Samsam Bakhtiari, Jean Laherrère, Kjell Aleklett, Chris Skrebowski, Jim Baldauf, James Howard Kunstler, Matt Savinar, Congressman Roscoe Bartlett, David Room, Jay Hanson, Richard Duncan, Walter Youngquist, Megan Quinn-Bachman, Barry Silverthorn, Julian Darley, Mark Robinowitz, and every other Peak Oil/Sustainability activist who has tilled the dry, hard, mostly unforgiving soil of human consciousness.

Special acknowledgment to Jenna Orkin, who today runs the blog that bears my name (http://www.mikeruppert.blogspot.com). You saved my life, and only you and I will ever appreciate the lengths you went to in order to achieve that. Your effort was not in vain. You and those who contribute to the blog are constantly improving our map and teaching more and more people around the world how to read it. You are giving the world a daily update on the progress of the collapse of industrial civilization and helping us to confront and mitigate it. Right behind Jenna stands the incredible soul and analytical talent of a man known to the world only as "Rice Farmer."

To all those who made From the Wilderness the world-changing, truly independent news organization that it was: Stan Goff,

Michael Kane, Carolyn Baker, Jamey Hecht, PhD, Jenna Orkin, Dale Allen Pfeiffer, and Dmitry Orlov.

A special acknowledgment to Stan Goff, a brilliant, courageous soldier with one heck of a dictionary, and to Mary, Kevin, and Patrick Tillman for helping to bring me back from the dead with a simple and sincere thank-you.

Special thanks to Margo Baldwin and all the staff at Chelsea Green Publishing, who stepped up at a critical moment and pulled me and this book back onto the right side of the looking glass. After all the tribulation of seeing this book come to life, you have proven Chelsea Green's commitment to quality publishing, service, and to leaving a better world for our descendants.

To Cynthia McKinney for her steadfast friendship and for teaching me how to find the strength to endure as long as there is breath in my lungs.

To James Michener, Ernest Hemingway, W. Somerset Maugham, and Henry Miller who, from my earliest years to the present moment, have given me faith in the power of well-chosen words.

To Larry Flynt for saving our First Amendment and for being such an inveterate rebel with a great sense of humor, willing and able to make fun of and challenge any sacred cow he sets his formidable eyes on.

To Cara Tompkins for being my test monkey with the final draft and becoming so important in helping me keep my act together throughout this long journey.

To my incredible attorney and dear friend, Wes Miller, who helped untangle the web holding this book back from the light.

To Doug Lewis, Andy Kravitz, and Squishy the dog for providing a place to check out of the struggle and find balance, laughter, friendship, creative energy, and precious music in our band The New White Trash.

To Emanuel Sferios and Paul Leddy, who have been such constant friends and who kept my Internet presence alive through very difficult challenges.

To Billy and Emma Kennedy, down under—the best mates a bloke like me could ever ask for.

To Chris Smith, Kate Noble, Barry Polterman, and Chris Thompson of Bluemark Films, who drew out and accurately portrayed the emotion and knowledge in me. It was you who have set the "real me" free, and I will never hide again. I salute your skill as filmmakers and your incredible ability to delve into and extract the essence of your character with loving ruthlessness. Bravo and thank you. *CoLLapse*, the movie, is not only a work of genius, but the most accurate rendition of my life and persona that I could have ever hoped for. You allowed Chelsea Green to use art from the movie, and that will make a huge difference in the number of people we can reach. You have provided a unique and priceless service to all mankind. History will rightly remember you for that.

A very special hug of gratitude to all of the great mapmakers at http://www.mikeruppert.blogspot.com, who have proved, and continue to prove, on a daily basis that the map we made is good; that it is teachable; and that the skill of mapmaking and -reading is transferable to those with open minds. God bless you all, my intellectual heirs and true family. It was you who—with your donations and gifts—kept me on the field to see this message delivered. This is not my moment in the sun. It belongs to all of us.

And to Rags the dawg, who is my constant companion, my psychiatrist, my spiritual advisor . . . and my best friend.

CHAPTER ONE

The Need for Leadership

If we have been lied to about mortgages, 401(k)s, stock portfolios, hedge funds, derivatives, insider trading, Ponzi schemes, appraised values, credit ratings, and adjustable rates; if we've been lied to by Bear Stearns and Lehman Brothers, AIG and Citigroup, Bernie Madoff and Stanford Financial; if we were lied to about the invasion of Iraq and torture; even about steroids in baseball—then why do so many accept on faith everything we have been sold about energy? Why accept it, especially when the people telling us about energy are the same folks who lied to us about everything else?

Why do people stridently defend a hyped-up, "no-problem" energy future based on promises of silver bullets which are accepted without the slightest bit of critical judgment or skepticism? When the energy information bubble bursts and the truth is finally known, it may be too late for our entire species to do anything about it. That will be the "bubble" that kills all of us.

This book hopes to prevent that outcome.

In 2009 the world has run short of energy—especially cheap, easy-to-find energy. Shortages, along with resulting price increases, have threatened industrialized civilization and the global economy. They actually endanger much more. It is safe to say that oil price spikes in June and July of 2008 broke the backs of over-extended consumers who could no longer meet their (sub-prime) mortgage payments and that this—and this alone—triggered the great economic crash which began in September and October. Energy and money are inextricably connected in very profound ways that this book will make simple and easy to understand.

Contrary to what the mainstream media say, there were many who actually predicted—in very stark and precise detail—the current economic collapse for years. I started issuing warnings in 2001, and current events have only confirmed what I and others saw coming. There were prophets who saw and warned of both the imminence and the ominous outlines of this collapse as far back as the late 1940s. Economic activity is not possible without energy, whether it be slave labor, horsepower, or oil. Starting in late 2001, just after the attacks of September 11, my staff of wonderfully talented writers and I started exploring this linkage. It was through understanding the connection between energy and money that we added great expanses of territory to a "map" we were making of how the world really worked, as opposed to features that were being increasingly exposed as lies, delusions, or mirages. Reading an accurate map enables one to see and understand past, present, and future. "If we keep sailing on this course, at this speed, the next point we reach will be . . . "

The current economic implosion will only result in the greatest and longest-lasting economic "depression" in human history—a new Dark Age—especially if some fundamental sea changes to the way we view both money and energy are not made immediately. We will start with energy.

The best way to understand all energy issues is to understand Peak Oil. The concept is simple.

Oil production has always followed a bell curve. Historical data compiled over a century of oil-production experience has established clearly and unequivocally that global production peaks approximately 40 years after discoveries do. Global oil discoveries peaked in the mid 1960s. According to Colin Campbell production peaks in individual countries usually follow peak discoveries after only around 25 years. This is due to technological advances in mapping and extraction. And keep in mind that all individual fields behave differently. Since oil comes from this planet, and this planet is, by definition, a closed system, Peak Oil means that once the top of the production bell curve is reached, no matter how much money, technology, prayer or marketing hype is applied; the planet cannot yield more oil in any following year—only less. And even less the year after that.

Oil is a non-renewable resource. This book will show clearly that of all hydrocarbon energy sources (oil, coal, natural gas and lesser sources), it is oil that is most important. It powers 90% of all transportation and is what plastics, pesticides, many chemicals, and a hundred other "indispensible" things are made from.

The edifice of human civilization, as it has functioned for a hundred years, is built upon cheap oil. There are four and half to five billion more people on the planet today than when oil was first used and that number is still growing. Those people, those souls, are here because oil, natural gas, and coal made it possible to feed, clothe, medicate, house, and transport them.

Oil geologist and physicist M. King Hubbert discovered and told us of this mathematical certainty in 1949.[1] Using his research and calculations he saw that oil discoveries in the United States had peaked in the 1930s. Based on three decades of data, he calculated that U.S. domestic oil production would peak in 1970. He was ridiculed and scorned by the scientific community. He was laughed at.

But he was absolutely right. Since 1970, U.S. oil production

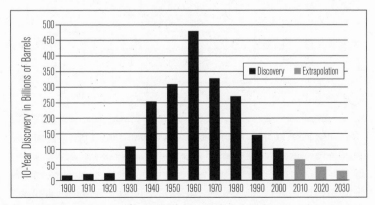

Oil discoveries have been declining since 1964. Note: World oil discovery over 10-yr periods, by Association for the Study of Peak Oil and Gas. SOURCE: THE OIL DRUM: HTTP://WWW.THEOILDRUM.COM/NODE/4172#MORE.

has been in a steady and irreversible decline. The United States was an oil exporter until about thirty years ago. Today the United States imports around 70% of the oil it needs to function on a daily basis.

Based upon currently available data, the production of conventional oil on the planet has peaked or is peaking now at around 86 million barrels per day. The low hanging fruit has been plucked. The biggest "orchards" with the cheapest-to-harvest "fruit" have been aggressively harvested for decades. No new apples have grown. (Oil, natural gas, and coal were produced by periods of intense global warming millions of years ago and "stored" until we discovered it). The orchards are now yielding less fruit every year. Yet our hunger and thirst continue to grow, even in the midst of an economic collapse where demand is shrinking. One reason for

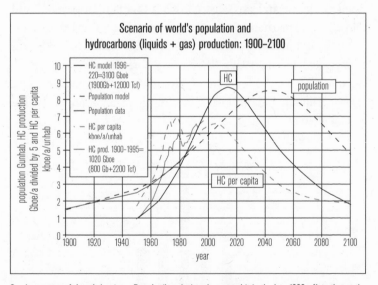

Graph, courtesy of Jean Laherrère a French oil geologist who created it in the late 1990s. Note that at the time, demand data suggested a peak in hydrocarbon energy production (oil, natural gas, and coal) sometime after 2010. There is an unbreakable correlation between economic growth and greenhouse gas emissions (the burning of fossil fuels). The current economic paradigm is based on this relationship. Note: Hydrocarbon energy production per capita peaked in 1979. FOUND IN *CROSSING THE RUBICON: THE DECLINE OF THE AMERICAN EMPIRE AT THE END OF THE AGE OF OIL* (2004) BY THE AUTHOR.

this is that the human population continues to grow. Another is that there is (according to several studies) a 96% correlation between economic growth and greenhouse-gas emissions. This is pure common sense. If there is to be *any* economic recovery, it cannot happen without spiking energy use again. What if gasoline prices go back to three dollars or three-fifty a gallon at a time of economic collapse; with a 10, 15, 20% unemployment rate; when there is no cash (or too much) in circulation?

Three-dollar gasoline in 2009 or 2010 will prove twice as deadly as four-dollar gasoline in July 2008.

These facts raise other, deeper, and more ominous, but utterly consistent, questions. What happens when natural gas enters decline? What happens when coal enters decline? What happens when fresh water runs low? What happens if population growth overshoots the ability of the planet to yield these "indispensible" commodities . . . like food?

Today there are on average ten calories of hydrocarbon energy in every calorie of food consumed in the industrialized world. That's before cooking energy is considered. Of the three, coal has had far less significance for agriculture. Take away the hydrocarbon energy and what happens? The food supply must shrink. Around the world nations are failing and people are starving because of energy and resource shortages. A dangerous game of musical chairs has begun. Other precious commodities, especially water and healthy soil (free from dependence on petrochemicals) are in increasingly short supply.

Corporations have discovered that they can continue to grow by helping human beings to fail or to become, as Henry Kissinger reportedly once said, "useless eaters." That is neither a rational nor moral choice for mankind. Over many years of study and research I have developed a map which shows that the U.S. (and world) economy is hopelessly corrupt and will behave that way. The accuracy of my "map" is part of what helped to place my first book *Crossing the Rubicon: The Decline of the American Empire at the End of the Age of Oil* in the Harvard Business School library and has kept it selling briskly four years after it was published. It ranked at or near Number One in Amazon's Public Affairs/Administration category for most of 2008.

In that book I detailed meticulously that the power elites, especially Dick Cheney (former CEO of oil-services giant Halliburton) and those he represents have known this crisis was coming for a long time. There is a plan to deal with the problem, and it has been kept secret from us. I believe the first obvious part of that plan was the invasion of Iraq which holds the second-largest known reserves of conventional oil on the planet.

A substantial part of the plan is in the records, minutes, and final report of 2001's National Energy Policy Development Group (NEPDG), which Dick Cheney chaired. That task force was funded with public money and Cheney fought twice, all the way to the Supreme Court, to keep its minutes and findings secret from the world and the American people who paid for it. We have a right to see those records, especially since oil prices increased around 500% between the attacks of September 11 and July of 2008.

The lay reader will likely ask, "Well, since oil prices have declined back to around $40 a barrel doesn't that mean we don't have to worry now?" The answer is, "Absolutely not. We should worry more." The short explanation is that no economic recovery (i.e. growth) is possible without using more energy. As long as oil prices are low we have stopped investment in possible oil substitutes, which are no longer profitable under the current economic paradigm. And, as this book will demonstrate, there is no combination of alternative energies anywhere—now or in the future—that will sustain the structure built by oil and fossil fuels.

The "map" many dedicated researchers and I constructed over a decade allows me to make very good predictions. The predictions I have made over the years have been spot-on when it comes to energy, money, war, geopolitics, and U.S. politics. The legendary Ted Williams only had to bat .400 to be baseball's best.

I have a track record that we estimate at better than .800 over the last eight years. There is a clear record supporting this claim.

Someone once said, "I would rather believe that there was a God and find out there wasn't, than not believe and find out there was." The great heroes of the Peak Oil/Sustainability movement have, for decades, been voices in the wilderness, warning about the biggest threat human civilization has ever faced. Can anyone argue that it is not better to realistically prepare for the energy crisis, which is here now, than to try and play catch-up later? We are tired of seeing our predictions come true year after year, and the disingenuous statements of Peak Oil deniers have worn so thin that we no longer see any need to debate the issue about whether this crisis is real or not. The good news and the bad news are that time is on our side now.

Pop psychology has made us all familiar with Elizabeth Kübler-Ross's five stages of grief. Indeed the human race will grieve over the loss of cheap energy and the way of life it brought. Those stages are: Denial, Anger, Bargaining, Depression, and Acceptance. Ross essentially said that acceptance meant, "I'm ready for what comes next." True leadership for America and the world must operate from the stage of acceptance. It cannot and must not tailor its approach to those who are in denial, who are angry or trying to bargain, or are depressed. That is the great failure of the media, politics and economics in the current paradigm.

Note that anger is the second stage. Already the world (and the mainstream media) is slowly and schizophrenically moving out of denial. When the anger stage arrives in full, there will be an enormous outpouring of rage, directed especially at the experts, media outlets, and snake-oil salesmen who told American consumers that excessive consumption and debt were good things; that there was plenty of energy, or that alternatives would permit the same levels of consumption that existed until the spring of 2008. Witness the planned and orchestrated collapse of the American housing markets, the wealth transfer taking place both out of the country and from public to private coffers. Witness the wholesale consignment of millions of people to financial ruin and debt-shackled poverty. That is just the beginning of what is to come.

Just a few short years ago, the United States government commissioned a study by Robert Hirsch of the defense contractor Science Applications International Corporation (or SAIC) to look at the problem of Peak Oil. When it was released in 2005, Hirsch concluded that "waiting until world oil production peaks before taking crash program action would leave the world with a significant liquid fuel deficit for more than two decades."[2] The implications of that simple reality are beyond the ability of most people to comprehend. This book will correct that.

Similarly, Professor David Goodstein, Vice Chancellor of Cal Tech, noted in his 2004 book *Out of Gas* that it takes 30 years to change an energy infrastructure.[3] That assumes that we know what infrastructure to change to, and that there would be hundreds of billions, if not trillions, of dollars and the raw materials available to do it.

Clearly, we have waited until the problem hit us in the face before taking action. And has anyone noticed that the United States is broke? We now have a national debt exceeding $11 trillion, and the budget deficit for 2009–2010 is expected to be over a trillion dollars.

Because I do not have to play "nice" with any political group or party, I am free to tell the truth, a truth dictated by easily understandable science and numbers. Because of my experience in and around government and elsewhere, I can take the reality that is now rocking human civilization and offer real suggestions that might make a positive difference in America and the world. This book, then, represents what the United States should be doing and saying rather than what it feels obligated to say on the world stage.

At this point I have an undeniable advantage. The feedback loops are now so short that we will know if this book was right or not within weeks, months or a year. I would bet heavy money that within a year (hopefully sooner) someone will be waving this book in the government's face and asking, "Why didn't you do this?" For the media and all those in suicidal denial about energy issues I am announcing right now that I will not go on the radio or TV to debate whether Peak Oil is real or not. Those who deny, or obfuscate, or mislead; those who argue that the human race can continue to behave and consume as it has for the last hundred-plus years will soon find themselves faced with the wrath of people who will understand that they have been misled.

The human paradigm has already shifted. As the legendary Colin Campbell, Peak Oil activist, author, and petroleum geologist has said for years, "The human race may not become extinct, but the subspecies of Petroleum Man almost certainly will."

The summer of 2008 saw dramatic price run-ups in the price of crude oil which, much worse than in previous years (always in the summer), had immediate and severe repercussions throughout the U.S. and global economies. Four-dollar gasoline caused a plunge in driven miles. It made it impossible for tens of thousands of homeowners to make their mortgage payments. To me and many more it looked as if the dreaded worst-case scenario of collapse might be underway with singular purpose. According to the media, a significant part of those price increases was attributable to two things: heightened tensions over a possible U.S. or Israeli attack on Iran and "speculation." They refused to focus on supply or depletion.

As prices fell by more than $30 a barrel in August and early September it became clear, however, that something else had been achieved. The markets and economic planners had found (and blasted through) the price-point at which demand could be destroyed. Regardless of optimistic or pessimistic outlook, all economists had recognized a singular issue as the planet went over the top of the Peak Oil curve. Called "the bumpy plateau" it has haunted economists for years. Essentially it says that at the peak of oil production there would be a series of relatively short spikes and dips where the price of oil would go so high as to instantly curb demand and then be followed by a marked price decrease which would spark more demand as consumers returned to old, unsustainable habits.

Witness the fact that over the 2008 holiday season most of those American consumers who bought new vehicles after huge incentives from the failing U.S. automakers bought pick-up trucks and SUVs.

But at some point after the bumpy plateau there was to be a cliff. That would be the cliff we are looking over right now.

At this writing, in 2009, the data suggests that once an economy is hobbled, oil-price decreases do little to increase demand. However, if there is to be any recovery it is axiomatic that more energy, especially oil, will be consumed. Though conservation of a non-renewable, irreplaceable resource is insufficient to remedy the current situation, it is a good thing when compared to the alternatives we will discuss in this book. It simply didn't have to be this way.

I was watching C-SPAN during the summer of 2008 and stopped for a second to watch debates in the House over offshore drilling. As the Republicans held the floor, presenting their unsupportable arguments that offshore drilling would produce enough oil in seven or ten years, there was a big sign behind the podium saying, "DRILL MORE, USE LESS." Clearly, reality was starting to sink in until the economy occupied the stage. What lawmakers and the media didn't get was that the economy was already the issue with high oil prices. Energy and money: Siamese twins.

But partisan arguing about thousands of undrilled offshore leases missed the point entirely. Those areas that have oil (except for formerly forbidden areas) are already being drilled. When oil companies obtained leases in the first decades of the 1900s, they took up almost every possible area where oil *might* be and then went back to explore later. The idea was to secure leases first to prevent competitors from getting them, and drill later if further study suggested there might be a find. Improvement in technology from the 1930s on ruled out many lease areas, but the leases remained in place in case oil rose in price sufficiently to justify drilling for small or difficult (i.e. expensive) pockets. Oil companies passed over the leased areas where it became clear that there was little likelihood of getting a sufficient payback for the heavy investment required. Democrats who argue that all untouched leases should be drilled seemed to be arguing that there is automatically oil there just because someone took an inexpensive lease seventy years ago. There isn't, and that's why the oil companies want to drill in the formerly prohibited areas where they know that at least some oil is to be found. We will look at what might be there later in this book.

The first rule in oil production is that there has to be oil in the ground before it can be extracted. Drilling holes does not produce oil.

Thus, by the time that both Barack Obama and John McCain had more fully articulated their energy positions with Obama promoting ethanol, and both sides promoting clean coal and other non-starters—the possibility of returning to old ways had been forever taken off the table. The economists had found the point at which they could destroy demand and buy a little time. Of course, no one talked about demand growth in China and India. The "recession" of 2008 seems to have addressed those issues as well. This is a global economic meltdown. No nation gets out unscathed. Nothing can grow forever.

The energy platforms of both candidates in the 2008 presidential election were hopelessly flawed, and this book will demonstrate that.

This policy will not be perfect. But I can guarantee that it will be vastly better than anything we will see from the White House or Congress. As the world-renowned energy author Richard Heinberg said to me recently, "After the politicians have 'dealt' with offshore oil drilling and oil speculators, which are only annoying gopher mounds, we will still have to deal with the Everest of Peak Oil." That and all the other energy crises that confront us.

No one else with better qualifications has stepped forward to tell us what the United States could or should be doing. I can at least say that I offered something here based upon reality; not upon the mind-numbing political and advertising babble which promises people things that cannot and will never be delivered. You will fully understand this by the time you finish this book, and it will change the way you live, the way you view the world and the decisions you make for yourself and your families.

The crushing need for this book became apparent when on June 5, 2008, a friend in politics asked me to be a vice presidential running mate on one of three non-mainstream party tickets. My first reaction was, of course, shock. My immediate second thought was that what was needed, especially as energy concerns had been monopolizing headlines, was a solid platform on energy. Nobody had one that made sense. John McCain and Barack Obama were both spouting "solutions" that were absolutely impossible, rife with fantasy and delusion. I declined the offer, but the desperate need for a platform—a real plan and an understandable policy— stayed firmly lodged in my mind.

I started making notes the next day.

I do not adhere to any political party or philosophy. All political parties in America are abysmal failures when it comes to energy and economics both. They cannot tell people bad news and they suppress those who try to. In fact, I believe that we are long overdue for a Jeffersonian approach; we should throw all archaic political constructs, buzz words, orthodoxies, and ideologies out the window and start with a fresh sheet of blank paper. We have entered a new human paradigm. It was and is of little benefit to keep refining Industrial Age government and philosophy in a new human paradigm.

The Democratic Party has become little more than an ineffective wing of one political construct, and its leadership in the face of the egregious offenses and fiscal irresponsibility of the Bush-Cheney years. It has become abysmally short-sighted, devoid of leadership or courage (with a few noteworthy exceptions). I was raised as a Republican in the days when the Republican Party meant something entirely different than it does today. I have never been impressed by Republican attitudes towards race and poverty. The Green Party is great on the environment but has historically been weak on people-issues, economics, and foreign policy. Libertarians have a decent understanding of money but no apparent grasp of energy issues. All the term "independent" means is that its adherents are sick of both sides. They have little new political ideology to offer. The American Left, especially the Progressive movement, is the most delusional, ineffective, and compromised political worm ball I have ever experienced, past and present.

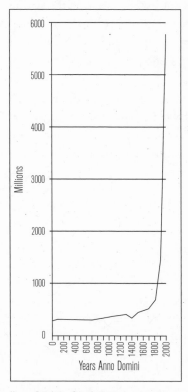

I view the entire American political and economic system as broken and corrupt, subservient to corporate/financial interests and an economic paradigm (based upon fiat currency, fractional reserve banking, and debt-based expansion) which demands infinite growth. That economic pyramid scheme—that mandate for infinite growth—is the beast which has driven us headlong into the unyielding steel wall of Peak Oil and the edge of the cliff lying just beyond.

I am not and will never be a candidate for any elected office. This policy, therefore, is written for the American people and their governments. I also have a distinct advantage

Human Population Growth. COURTESY COLIN CAMPBELL.

in that no one else has one. I have no real competition. I will make errors. What I write here can most certainly be improved upon, and I expect that. This book, however, will be a resounding success if it opens the door to real dialogue on energy and money. One cannot be discussed without the other. An honest public discussion on these subjects has never taken place.

In this policy statement the reader will find elements that are philosophically akin to FDR's New Deal and some positions that sound like they are straight out of a conservative Republican or Libertarian playbook. I have only one guiding principle: to implement policies that will keep the nation functioning and that will protect the American people and the world as a whole. No government should protect corporations and banks with secret distributions of taxpayer money while allowing the nation as a whole to fail. This is a practice that is horribly short-sighted and destructive of the one thing that the American people will need as the Siamese twins of the energy and economic crises determine our future—options.

There are, in this book, many recommendations that can be taken by individuals, families, and communities without relying on government. A thorough reading of what follows will make that clear and hopefully disclose other steps I have not considered to the discerning reader.

CHAPTER TWO

Thinking Like a President in the Face of a Global Problem

Energy, not money, is the root of all economic activity. Money represents only the *ability* to do work. By itself a dollar bill can do nothing. You cannot put it in your gas tank and expect your car to run. Energy is that which money symbolizes, whether it is the slave labor of centuries past which built civilizations that later perished, the food that comes to your table today, or the gasoline that goes into your car or the electricity that comes into your home. Cheap energy has always been the equivalent of free slave labor for industrial civilization.

There is one other essential difference between money and energy. Money can grow infinitely. Energy, i.e. the slaves necessary to give money value, cannot.

What happens when the "slaves" disappear?

In the first decade of the twenty-first century it has become clear that a major crisis confronts the human race. For the United States it will be, and is, a crisis as great as any we have ever faced as a nation. It will be greater than World War II, greater than the Civil War, and much more devastating to American life than the Great Depression of the 1930s.

This crisis will be different from all others that came before it. In the past America faced all of its great challenges by marshalling and applying our ingenuity to use the one advantage America had over all other nations, our vast and largely untapped base of natural resources. In the Great Depression, America responded with a New Deal. It drew upon seemingly endless reserves of energy and raw materials to build its way out of economic collapse with massive public works projects and preparation for war. In World War II the Manhattan Project took advantage of seemingly unlimited reserves of energy, hydroelectric power, and minerals to enrich uranium and unleash the power of the atom. When President John Kennedy challenged America to put men on the moon and bring them home safely by the end of the 1960s, every commodity needed to achieve that task was cheap, plentiful, and close at hand.

We live in a different world now. The crisis we face is itself a crisis of shortages (not just energy), and if America is to meet the test successfully, we will have to use a different approach. It is not possible to use enormous amount of resources to address a resource shortage.

In our two greatest historical challenges, the Civil War and the Great Depression, Presidents Abraham Lincoln—a Republican, and Franklin D. Roosevelt—a Democrat, found it necessary to exceed the powers granted them by our Constitution. History has judged that both men acted wisely, and it has well recorded their emotional agony as they gave their health and their lives to protect the Union they loved so dearly.

Barack Obama and the presidents who follow him will be faced with a greater crisis, requiring decisions which will try their minds and souls as no other challenge has tried an American president. They will have to make choices which will be difficult and cannot please all of the people all of the time. They will have to set priorities which cannot give everyone an equal seat at our great table. They will have to eject some from long-held seats to make room for new ones who can help America and mankind move to a new post-industrial paradigm. And they will have to do all of this holding one standard above all others: do what is best for the American people as a united whole. Do what is best for mankind.

What is happening is not just an American crisis. It is a global crisis. It is an emergency for every human being alive today and especially for those as yet unborn. That crisis is a combination of two factors which are converging like a giant claw around all of us. One half of the claw is an exponentially surging rise in human population and its need for food, energy, goods and services. The other half is a rapidly declining supply of cheap, affordable energy in all its forms with which to manufacture, transport and power those things, whether they are automobiles, TV sets, cell phones, clothing, computers—or, most especially, food.

This is not a Conservative or Liberal issue. The high cost of gasoline, food, electricity, and everything else today impacts everyone, regardless of their beliefs. This policy is not designed to reflect any political philosophy. In fact it is an argument that all pre-existing political philosophies be thrown out. This policy is based upon science: mathematics (in most cases high school- or middle school-level mathematics), geology, and simple chemistry. Numbers are impersonal and apolitical. They can be—and especially when it comes to energy—have been manipulated and misused both intentionally and ignorantly. There is too little clarity and too much confusion when it comes to how much energy there is and what alternative sources are available—and what we can realistically expect to accomplish with them.

Yet one thing is frighteningly clear to everyone regardless of their beliefs: the Age of Oil is coming to an end. Whether you are someone who believes that the United States must somehow

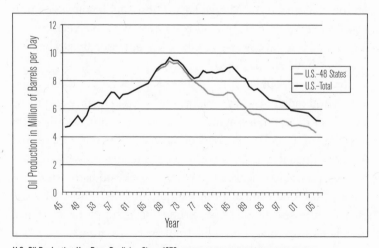

U.S. Oil Production Has Been Declining Since 1970. SOURCE: US ENERGY INFORMATION ADMINISTRATION.

wean itself from dependency on foreign oil or you are someone who understands that the human race has failed to plan and prepare for this crisis in time to make a stable transition to some other regime (or combination of regimes), the objective and the urgency remain the same.

Domestic oil production in the United States peaked in 1970 and has been in an irreversible decline since. Today the United States imports more than 70% of its oil from foreign sources, and no amount of domestic drilling will narrow this ever-widening gap.

This policy is also based upon the global political and economic realities of the first decade of the twenty-first century. Let us begin with facts that are not in dispute.

The first important reality is that the United States of America represents 5% of the world's population yet it uses 25% of the world's energy. Therefore, since the United States holds itself up to the world as a leader, an innovator, a champion of human rights and equality, it must lead by example. American energy policy cannot be created in a vacuum. All countries in the world, including our own, depend upon oil that, in almost every instance, comes from some other country. 60% of the known oil on the planet is in the Middle East. Those supplies are depleting rapidly, even as global demand is soaring. We have been pumping them flat-out for nearly 70 years. It is unreasonable to believe that they have not declined.

Contrary to popular fantasy, there is not a lot of oil left to be found; and much of what remains is of a lesser quality and more expensive to refine, in smaller reservoirs, or simply not oil as we have known it. A sizeable portion of that "oil" cannot be turned into gasoline. It can be used, however, to make asphalt, some plastics, pharmaceuticals, or to power a backyard barbecue.

As with money, one cannot put "technology" into a car's gas tank or into the tanks of an airliner or a ship. One must put petroleum into those tanks. Technology comes from energy, not the other way around. The laws of thermodynamics dictate this.

A little over four decades ago, as discoveries of oil peaked around the world, the human race was finding roughly as much new oil as it used each year. Since then, global oil demand has risen from roughly 50 million barrels per day to more than 85 million barrels per day. In the meantime, the discovery of new reserves has dwindled to one barrel of oil (or non-conventional oil) discovered for each four to five barrels used, and that gap is widening rapidly. One study by a leading expert from Sweden predicts that by 2030 the world will be using 10 barrels for every new barrel discovered.[1] By that time it may cost $100, $200, or more per barrel to extract what oil remains.

The world will never run out of oil. Once it takes more than the sale price to extract one barrel, or it take more energy to extract a barrel than one gets from burning it, there is no point in using it.

The term "non-conventional oil" is important. Because it is through understanding that term that we begin to look at how much energy we expend for how much energy we get in return. Unconventional oil sources like Canadian tar sands or oil-from-coal—which has not been proven to be commercially viable and is very destructive of the environment—put us face to face with the fact that nothing will ever provide an energy return for energy invested like the oil we began pumping at the beginning of the last century. Nature made that oil over millions of years. There is no free lunch. The Laws of Thermodynamics are as fixed as the law of gravity.

Demand growth has only slowed, not reversed, and any "recovery" will only require more. No alternative energy regimes exist that can replace current levels of consumption, let alone those that are coming as a result of population growth in other parts of the world.

There are roughly 800 million internal-combustion-powered vehicles on the planet: cars, trucks, ships, and planes. According to a number of sources, there are more than 250 million passenger vehicles in the United States alone. Not only do they run on oil and emit greenhouse gasses, they are literally made from it. According to the *National Geographic*, there are seven gallons of oil in every tire.[2] All plastics and vinyl are made from oil. Oil is a vital component of paints, resins, and adhesives used in automobile manufacture. And enormous quantities of oil, natural gas, and coal are required to construct a vehicle from raw materials. So even if it became possible to replace all 800 million vehicles with some new kind of power source, the world would have to use up much of the oil that remains to make these new vehicles powered by technologies that aren't (and will never be) ready in plant assembly lines that do not exist and would have to be constructed using, coal oil and natural gas.

Myths of hydrogen-powered vehicles are just that, myths. According to a University of California study, it takes 1,113 gallons of gaseous hydrogen to equal the energy in one gallon of gasoline.[3] Commercial hydrogen today is made from natural gas, and natural gas is also in short supply.

Electric cars sound nice, but electricity is not an energy source. It must be generated from energy. The idea of using fresh water as a fuel source, where electricity is inefficiently used to split water atoms and release hydrogen, is not rational. The world—especially as a result of climate change—is running out of water for drinking and irrigation. Drought is ravaging much of the planet. Glaciers are melting. Even the most ardent promoters of hydrogen technology agree that commercially viable hydrogen vehicles are not possible for at least three decades, and that is only on the wishful assumption the certain technological "break-throughs" take place in the meantime. These breakthroughs can never overcome hydrogen's weaknesses. Our crisis is now. Political leaders cannot gamble the future of this country on wishful thinking about what might be possible. If only. If only.

New cars and trucks will, if they are ever made, all be powered by oil.

A great many people around the world, not just in the United States, are not going to have their expectations met. The rest of the planet will not sit idly by and watch the United States consume what is left. Contrary to what Vice President Richard Cheney said defiantly a few days after the attacks of September 11, the American way of life *is* negotiable . . . and breakable. Americans no longer live in a world where rampant and wasteful consumption is an economic mandate.

This is the first thing all public officials must understand.

CHAPTER THREE

Depletion: Refilling Niagara Falls with a Garden Hose

Depletion of existing oil reserves is both pronounced and accelerating. In 2005, it was reported that 33 of the largest 48 oil-producing countries had entered decline.[1] Data compiled in 2008 showed that of the 50 largest oil-producing countries in the world 42 had passed their peak and are in decline.[2] In other words, nine more major oil-producing countries passed their production peaks in the last three years alone. Indonesia, a founding member of OPEC is now importing oil to meet its domestic needs. Major oil exporters like Mexico and Kuwait are experiencing a geologic event known as oil-field collapse. In 2005, the world's second largest oil field, Burgan, in Kuwait, collapsed[3]—producing a dramatic decline in production or "exhaustion" where production rates don't softly decline—they plummet. This is a result of

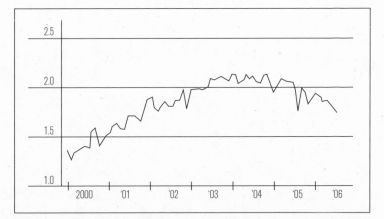

Over the Hill? Output at Mexico's huge Cantarell oil field is falling faster than expected, a worrisome development for both the country and energy markets. Monthly data, in millions of barrels per day. SOURCE: MEXICO'S ENERGY MINISTRY.

heavy water or nitrogen injection and the resulting collapse of geologic structures this often causes. In Mexico, the world's third largest oil field, Cantarell, has experienced the same fate and is also in rapid decline.[4]

Mexico has historically been America's second-largest oil supplier, and it is clear that it cannot hold that position much longer. As in many other "developing," oil-exporting countries, each year brings a demand that more and more oil be kept in the source country for domestic consumption to avoid civil unrest. Mexico's overall output fell 11% the year ending June 2008 (when demand and prices were at their highest), and the output from Cantarell fell by 35%.[5]

It is the collapse of Mexico's oil revenue, not drug wars, which has made that country into a failing state.

Contrary to popular belief, not all oil is recoverable. Once it takes more than one barrel of oil or other energy equivalent to extract one barrel of oil, an oil field is considered dead. What's the point? Would you spend $10 for a nine dollar return? Would you burn one barrel of oil to get .9 barrels back?

Britain's once-prolific North Sea fields are near complete exhaustion (for oil). Norway, once one of the largest exporters in the world, has seriously declined. For more than a decade the Norwegian government has been on a crash program preparing for a post-petroleum world. Russia, the world's second-largest producer, is in decline. Iran is in decline. Kuwait and Nigeria are in decline. Venezuela is in decline for conventional oil production while its heavy-oil deposits in the Orinoco Basin awaiting new technologies and the construction of very expensive refineries.

Regardless of what Americans think about Venezuela's leadership, that country has been America's fourth- or fifth-largest supplier of oil for decades.[6]

Alaskan production is down to 37% of what it was at its peak in 1988.

This arithmetic of depletion is not as vague as it is for honestly

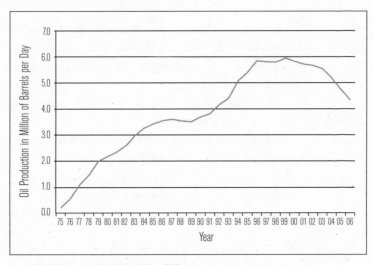

North Seas Oil Production Has Declined Since 1999. SOURCE: US ENERGY INFORMATION ADMINISTRATION.

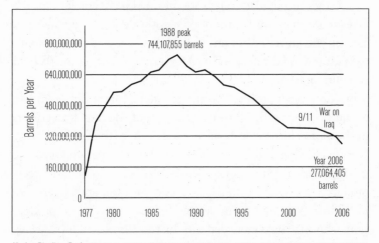

Alaska Pipeline Peak. POSTED AT WWW.OILEMPIRE.US/ALASKA.HTML. DATA SOURCE: WWW.ALYESKA-PIPE.COM/PIPELINEFACTS/
THROUGHPUT.HTML.

estimating recoverable reserves still in the ground. Decline is
decline, and according to various estimates, world oil production
is declining at somewhere between 4–6% per year. And most
importantly, that decline rate is increasing. In early 2008 the
International Energy Agency's chief economist, Fatih Birol, esti-

mated decline at 5.8%. That means that if Planet Earth produces 84 or 85 million barrels per day in 2008, it will be able to produce only around 80 million barrels per day in 2009. Many experts I interviewed for this book believed the decline rate is actually much greater, and their assessment did not have to wait long for validation. In September, 2008, I attended an Association for the Study of Peak Oil—USA (ASPO-USA) conference in Sacramento, and a source within the oil industry told me that a major oil-services company had just pegged the global decline rate at near 9%.

By late 2008 the International Energy Agency confirmed this with an acknowledged global decline rate of 9.1%. Actual decline rates will be lower because of new production, but it is certain that we will still be in serious decline no matter what. And it is this irreversible decline that threatens human civilization. It does not matter if there is 5% less oil next year or 9%. That report was leaked to the *Financial Times* but was not the subject of an emergency press conference as it should have been.[7]

Just as the feature documentary film *CoLLapse* featuring me was receiving most-favorable reviews in American and Canadian mainstream press in November, 2009, a bombshell was dropped by two highly-placed whistleblowers from within the IEA that served to virtually destroy any credibility the agency had in terms of reserve estimates and decline rates. A Nov. 9, 2009, story in Britain's *The Guardian* caught the world's attention with its headline: "Key oil figures were distorted by U.S. pressure, says whistleblower". The first two paragraphs of the story essentially confirmed what most of us in the Peak Oil movement had been saying for years: IEA numbers couldn't be trusted. The Peak Oil movement had been relying on raw production data rather than IEA administrative summaries prepared for a mainstream press with no appetite for bad news.

The world is much closer to running out of oil than official estimates admit, according to a whistleblower at the

International Energy Agency who claims it has been deliberately underplaying a looming shortage for fear of trigger-
ing panic buying.

The senior official claims that the US has played an
influential role in encouraging the watchdog to underplay
the rate of decline from existing oil fields while overplaying
the chance of finding new reserves.[8]

This "revelation" reminded me of Claude Rains's mock surprise
in the film *Casablanca* when he proclaimed loudly to Humphrey
Bogart, "I'm shocked! There's gambling going on in this establishment." But it was no surprise for the Peak Oil movement
and experts like Colin Campbell, Richard Heinberg, and Matt
Simmons and so many others who had been saying for years
that IEA's books were cooked. We had all been saying that the
IEA was about as trustworthy a watchdog as the Securities and
Exchange Commission had been in protecting people from the
cooked books of Bernie Madoff, Lehman Brothers, Fannie Mae,
AIG, and Citigroup.

The whole economic paradigm is predicated upon the need
for more oil every year. The imperative has always been to keep
people spending and consuming instead of dealing with reality.
That information and what it means was not publicly acknowledged anywhere by any individual or entity to which people have
been conditioned to look for answers. Hence, it was not seen.
The 9.1% figure now means that in 2009 it would not be possible
to produce more than perhaps 80 million barrels per day, the
only mitigating factor being new fields coming online with much
lower production rates or technology which is not that new and
only minimally effective in increasing production. All of this
assumes that that oil returns to $90 or $100 per barrel to make
new and smaller fields, or the cost of technological investment
profitable.

But, as we will see at the end of this book, the economic
collapse begin in 2008 has been so severe that even $80 oil is

proving to be a demand limit. What that means is that fields that need $90 or $100 oil to justify development will sit idle.

New discoveries usually take between four and seven years to bring online. So what "new" oil is coming between 2009 and 2014 is already a fairly well-known quantity. As we shall soon see, what's coming online now is just "a drop in the bucket." It is not enough to satisfy even current (2009) demand, let alone any demand growth that an economic "recovery" would require. New oil coming online is not even close to making a dent in depletion! And what that means is that big price spikes lie in wait to ambush a greatly weakened global economy.

So let the "Drill Baby Drill" (very expensive drilling) nincompoops get in line for their own Darwinian deselection contest. Some of us have chosen not to drink that Kool-Aid and are looking for alternatives.

At the old (deliberately deceptive) decline rate of 5.8%, almost five million barrels per day in new production would have had to come online in 2009 just to offset decline. Assuming a constant rate of decline (which is not how decline works) another 4.2 million barrels per day of new production would have had to come online in 2010. All of the new drilling in the U.S. coastal regions and Alaska will not come online until about 2014, and no one, anywhere, has predicted that kind of production. In the meantime America's other fields will continue to produce less and less oil.

At the new decline rate of 9% we would need roughly eight million barrels a day of new production to offset decline. As one of the world's leading energy experts—who happens to also be the world's largest energy investment banker—Matthew Simmons, has said repeatedly to me and in countless lectures, "We need to find three new Saudi Arabias just to offset decline."

They aren't there to find and we are just a tad behind the five-to-seven year development phase to make them produce in time for an economic recovery . . . if they existed.

Prior to the economic collapse of 2008 the International

Energy Agency and OPEC had estimated a 30% global increase in demand for oil over the next two decades. However since late September of 2008 the media has been flooded with stories about how the growth in demand has reversed. As of September 2009, the per barrel price hovers near $75–80. Is that good news? Not for the unemployed, homeless, and all nations where GDP is shrinking. They no longer have money to buy gas at any price . . . or anything else for that matter. And what they don't buy doesn't get shipped or replaced so oil demand evaporates. Demand was not curtailed by $147 prices. It was destroyed. That's why I refer to energy and money as Siamese twins. In the current economic paradigm they cannot be separated without killing both.

The largest oil field ever found, Saudi Arabia's super-giant Ghawar, discovered in the late 1940s, has been estimated to have contained just around 100 billion barrels of oil. Humans now use 1 billion barrels of oil every 11–12 days. Sixty years after its discovery, geologists are unsure as to how much oil was originally there because it is impossible to measure exactly and very difficult to actually estimate, even when hundreds of wells are in place and producing. Contrast this with announcements in the press that, based upon one test well for example, a new field off the Brazilian coast may have 33 billion barrels.[9] The Brazil story represents dozens of unsupported press releases and announcements in recent years from all over the world, reported as fact, that caused people to invest, to buy bigger homes, and to consume more products made with and by oil and natural gas. They were all lies.

However, major American financial institutions like Goldman Sachs and the World Bank recognized around 2004 that when there is no supply left to expand, the only option is to destroy demand. In 2005 a Goldman Sachs report observed (as oil reached $57 a barrel), "Ultimately, we agree that the energy bull market will roll over once demand destruction really begins. We simply do not believe we have arrived at that point."[10]

In his seminal report on the challenges of Peak Oil (commissioned by the U.S. government), Robert Hirsch of SAIC wrote:

> If mitigation is too little, too late, world supply/demand
> balance will have to be achieved through massive demand
> destruction (shortages), which would translate to extreme
> economic hardship.[11]

Is any of this sounding familiar?

That there is a universally accepted, near-perfect correlation
between greenhouse-gas emissions and GDP growth means that
the financial "powers that be" and the Bush administration
decided (or accepted) that the *only* (under the current paradigm) thing to do is to destroy economic activity—to shut down
businesses, to put people out of work, and to take food from our
tables. Is that not a process that the American people should
have a say in? Is that not a dialogue that every human should
have a voice in?

In the early and mid 1990s American and world media trum-
peted that the Caspian Basin in Asia would have 250–300 billion
barrels of oil. Now, almost twenty years later, those rosy estimates,
which once sent stocks and consumption soaring, have been
reduced to 30 to 40 billion barrels of recoverable, heavy-sour oil.
Twenty of the first twenty-five wells drilled in the Caspian basin
were dry holes. The first well alone cost $125 million.[12]

Where is the world going to find that kind of money now? All
the new money being printed out of thin air is going to service a
$700 trillion derivatives bubble to keep banks and lending insti-
tutions afloat. All that money is doing is enabling the financials
to try and make their minimum monthly payments on a credit
bubble they created.

Prior to the economic collapse Arctic oil was a hot topic. It
will take a minimum of ten years to bring online, and because
it is all underwater, it will never produce as rapidly as Ghawar,
which has produced as much as five-plus million barrels per

day. The Arctic does not belong to the United States. Russia, Canada, Norway, Sweden, and Greenland all have legitimate claims there. Arguably, the entire human race has a claim there. What's more, these Arctic optimists neglect to state that they view the complete melting of the polar ice cap as a good thing. I and many polar bears disagree with that.

The problem with the polar ice cap is that is moves. One can't put a rig on it on Tuesday and expect it to be over the same spot on Thursday. What the heck . . . let's just melt the damn thing so we can put a $200 million rig there that may or may not find oil.

But following the economic collapse, with oil at around $75 there isn't (pardon the pun) a snowball's chance that Arctic oil will get developed at all. Why spend between $90 and $100 or more a barrel to get oil that you can only sell for $75? Remember that under the current economic paradigm, the four- to seven-year development clock doesn't start until the oil price becomes profitable. In the meantime all the other big reservoirs are declining at an accelerating pace.

The gaps between our energy visions and energy realities are widening.

No field even close to the size of Ghawar has been discovered since 1948. According to Matt Simmons, we need to find those three new Ghawars just to offset decline, let alone accommodate anticipated 20–30% growth in demand. Decline will prevent the planet from just getting back to its consumption levels of July 2008 and that is where both Washington and Wall Street fall flat on their face.

If Ghawar is likened to an Olympic-sized swimming pool which holds approximately 600,000 gallons of water, the Arctic National Wildlife Refuge (ANWR) in Alaska, optimistically estimated by many to hold between five and 12 billion barrels of recoverable oil in widely scattered deposits, is about the size of one average-sized backyard swimming pool holding 30,000 gallons—or 5% of Ghawar.

There are no roads there. There are no wells there. There is no pipeline in place. The necessary infrastructure is enormously expensive, and there is no guarantee that that much oil will either be there or be recoverable if it is. Remember all those glorious predictions about the Caspian? Did you or your family go out and buy an Escalade or Expedition after you heard that news? How many 125 million-dollar wells will we be able to pay for? How many will be dry holes?

As the brilliant energy expert and columnist Tom Whipple said to me in September 2008, "What drilling in ANWR means is that in the year 2016 or 2018, Americans will be paying $11 a gallon instead of $12, if the economy holds out. And that will only last for a short time."

The rush to drill off America's coasts is a rush to find small swimming pools, hot tubs and even bathtubs of oil. It may be necessary to do that but not with the expectation of a return to lower prices and the same consumption patterns of years gone by. Those deposits will be needed to save lives and help maintain basic services. It is short-sighted of all the environmentalists to oppose all new drilling on federal lands unequivocally.

Should any American leader or Congress or state legislature gamble with the future like this, especially when—as we shall soon see—reserve estimates from governments, oil companies, and market analysts have long been shown to be totally unreliable; estimates prepared for Wall Street, for tax purposes, to buoy consumer confidence and keep us spending rather than facing the truth?

Natural Gas
Natural gas is another critical energy source for modern civilization. About one half of America's electrical generation comes from natural gas-fired plants and more are being built. It was made from the same organic materials as oil, only "cooked" and concentrated for millions of years at slightly different temperatures and pressures. Natural gas differs from oil in two significant ways. First, unlike oil production which generally follows a

smooth bell-curve of production, it is a gas which just blows until the pressure drops and then it shuts down like the air leaving a balloon. This is called the Natural Gas Cliff.

Second, natural gas cannot be imported the way oil is. As it is now, natural gas must come to the United States by pipeline, either from Canada, Mexico, or the Gulf of Mexico. United States domestic production of natural gas has been falling (with recent exceptions, see Chapter 9) and there is a clear trend showing that all the large domestic deposits have been discovered and used. So-called shale gas is a new development but shrouded in a great deal of secrecy about its methodology, toxicity and fresh-water consumption. We will discuss it and show that it may not be the reprieve we hope for or have been sold through press releases that get read on the air or quoted without any critical analysis.

In order to be shipped between continents, natural gas must be compressed into highly explosive liquefied natural gas or LNG. LNG tankers and terminals are enormously expensive to construct and pose huge safety risks, as well as very tempting targets for terrorists.

In the United States, natural gas currently performs two vital functions. It is responsible for roughly 40% of our electrical generation and it is the feedstock for all nitrogen-based fertilizers with which we grow food. Natural gas is also the feedstock for all commercial hydrogen and many vital chemicals. In Canada, enormous amounts of natural gas are now being burned to generate steam to wash heavy oil from tar sands to power American vehicles. Suggestions that we convert to natural gas-powered vehicles only push us closer to that unthinkable future.

The largest known reserves of conventional natural gas in the world are in Russia and Turkmenistan. They do the United States no good and only offer us the choice of switching our dependence from one set of countries to another, even if the LNG infrastructure, which would cost hundreds of billions, if not trillions, of dollars were already in place.

Just a few years ago the United Kingdom became dependent upon Russian natural gas to keep its citizens warm in winter. Britain then surrendered its energy sovereignty to the European Union as I had predicted it would as far back as 2002.[13] The reason: all the pipelines carrying Russian gas to the UK flow through the heart of Europe. The cost of heat there has become so expensive that each year as many as 50,000 British citizens, mostly elderly, freeze to death because they have to choose between food and warmth.[14] When Russia has an especially cold winter, or has decided to flex its muscles in the foreign policy arena, European thermostats get turned down.[15]

In recent years Russia has twice shut off gas flows through Ukraine, which plunged parts of Europe into crisis management to keep its citizens from freezing and (here's the money connection again) to keep factories open.

What options do we have in the United States?

CHAPTER FOUR

Reserve Estimates: Playing a Fool's Game with Numbers

From the International Energy Agency, to the U.S. Department of Energy to Wall Street to oil companies themselves, it is universally agreed—not just in the U.S. but around the world—that there is no transparency with regards to how much oil is in a field. It is, from a geologic standpoint alone, very difficult to actually determine. When you throw in manipulation, economics and politics the waters become almost impenetrable.

In short, no one in any government anywhere completely trusts any reserve numbers published by any oil company or any oil-producing nation. In spite of that, major media outlets around the world routinely trumpet inflated and unreliable numbers in thoughtless sound bites. The American people then accept those numbers and say or think that there is nothing to worry about while they are maxing out their credit cards, reinforcing their consumption patterns (a benefit for Wall Street). Even most members of congress don't have a clue when they talk about reserves; and those who do only say what their campaign donors tell them to. Perhaps they suffer from the politician's disease of being unable to tell constituents any bad news at all.

Fatih Birol, Chief Economist for the International Energy Agency, said in a 2008 interview with *The Energy Bulletin*:

> **BIROL:** We are talking about a very important issue here, and the most important accomplishment I expect from the WEO [World Energy Outlook] 2008 is more transparency as far as the oil reserves of the national as well as the international oil corporations are concerned.

SCHNEIDER: Who are you hinting at?

BIROL: Just remember that a very well known international oil company has recently run into trouble because it did not have enough transparency. Therefore the IEA would like to see more openness in accord to data about oil reserves—it might be the national good of the individual states, but the rest of the world, other economies, the common wellbeing of everyone are dependent on it. At the moment we are flying almost blindly and we desperately need more insight here.[1]

Dr. Colin Campbell, a senior oil geologist, retired oil executive and one of the most respected experts on Peak Oil wrote to this author recently about his new book (*The Atlas of Oil and Gas Depletion*), in which he was able to produce a reliable statistical picture of depletion patterns around the globe. He said that when it came to reserve numbers as presented by companies and nations, "the only numbers that are certain are the page numbers."

Companies and nations have never had to, or been accused of, hiding numbers that were larger than expected.

Do you know the difference between estimated reserves, probable reserves, proven reserves and ultimately recoverable reserves? They are accounting creations cooked up to value share prices and borrow money or attract investors. They have nothing to do with how much oil is in the ground. I have seen these numbers vary by as much as 300% for one field or region. Certainly the American media does not explain this. The truth about reserve numbers is that they are ledger entries more than honest scientific analysis. Oil companies have to pay taxes on reserves so they use smaller number when reporting those. But when it comes to reporting to stockholders and the media, they use larger numbers to encourage consumers, boost the markets and inflate their stock price.

Royal Dutch Shell, one of the world's largest companies, was

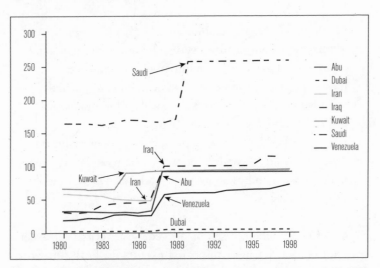

Falsely "never-emptying" reserves and huge jumps in reported reserves during "quota wars" when OPEC allowed exports (and therefore income) according to a country's reported size of resources. DATA ARE FROM PETROCONSULTANTS OF GENEVA, A CONSULTANCY WHOSE DATABASE IS THE MOST COMPREHENSIVE AVAILABLE FOR DATA ON OIL RESOURCES THAT EXIST OUTSIDE OF CONTINENTAL NORTH AMERICA, AND IS USED AS A 'BIBLE' BY ALL INTERNATIONAL OIL COMPANIES. (PREVIOUSLY PUBLISHED IN *CROSSING THE RUBICON: THE DECLINE OF THE AMERICAN EMPIRE AT THE END OF THAT AGE OF OIL* BY THE AUTHOR).

caught falsifying their reserve numbers in 2003 and 2004. They had to downgrade their reserve estimates not once, but four times, and were penalized for it. The two co-chairmen of Shell (one from Britain and one from the Netherlands) were forced to resign, and the scandal triggered a wave of reserve restatements throughout the industry. [2]

Nations are no different, especially within OPEC, where production quotas are set as a fraction of "proven" reserves. In the mid-1980s OPEC wanted to produce a lot more oil, but their quota system prevented that. So what did they do? Every OPEC member except Dubai broke out their erasers and voila! As a result, OPEC members could produce and sell a lot more oil. Cheap oil flooded the markets, and the Soviet Union, which depended on foreign currency, went bankrupt..It is impossible to believe that Mother Earth just refilled everyone's reserves and somehow forgot Dubai.

There is no universal standard for reserve reporting or verification, and this has allowed both nations and companies to play with the numbers to suit their own ends.

The utmost priority for all nations is to know, with the highest-possible degree of certainty, how much oil is left, where it is, and what kind it is. Is it Texas midgrade? Is it Brent crude? Is it heavy-sour and full of sulfur like the oil from the Caspian basin. All of that determines how expensive it is to refine and the number of refineries that can handle it.

Much of that work has already been done, but it has been kept a secret from the American people and the world. That information is contained in the report and minutes of Vice President Dick Cheney's National Energy Policy Development Group (NEPDG), which finished its work just months before the attacks of September 11, 2001, and then fought all the way to the Supreme Court to conceal what they found.

For a nation like Saudi Arabia, holding about 25% of the recoverable oil on the planet, actual reserve numbers are—as with almost all other countries having nationalized oil companies—a closely held state secret. The Saudis have been pumping their oil virtually flat-out for six decades. However, if one looks at their currently declared reserve numbers, one finds that the Saudis are claiming to have almost the same amount of oil as when they started. Imagine that.

Saudi Arabia has never produced more than about 9.5 million barrels of conventional oil per day. Media pundits both ignorantly or deceptively throw out numbers like 11 or 12 billion barrels per day for future Saudi production and include natural gas liquids which are useless to make gasoline but great for your propane or butane tanks—or extra heavy-sour oil, which is fine for making asphalt and often can't be refined into gasoline. Many experts, including Matthew Simmons, the world's largest energy investment banker, (as he documented in his best-selling book *Twilight in the Desert*), believe that Saudi Arabia may well have entered decline and be on the brink of collapse in some fields—including the great Ghawar.

After years of study I agree that Saudi Arabia is hiding a great deal.

The implications of this are dire as the world hopes for an economic recovery from the crash of 2008–2009, both for the world and for the Saudi monarchy—a monarchy that was created by Western powers in the 1920s and 1930s strictly for the purpose of securing Saudi oil for the west.

First, it is automatically a given that if Saudi Arabia has entered decline, the planet Earth has entered decline. Oil geologists have been scouring the planet since the 1920s. It has never been terribly difficult to locate what petroleum geologists call "source rock"— the kind of rock, at the rights depths, which *might possibly* hold oil. Over the decades, exploration technology has improved dramatically, allowing geologists to find the right source rocks and make predictions to varying degrees of certainty as to how much oil might be there. The *how much* might be there is basically a factor of how large an area the source rock covers and other geologic data, not *how deep* it is. Source rock does not guarantee the presence of anything except source rock. If hydrocarbons were deposited 150 million years ago at too shallow a depth in the source rock, one *might* find natural gas. Go too deep and one *might* find coal.

One never knows what will be found until one drills. So there is no chance of finding another Ghawar, let alone the three or four that are needed to offset decline in other countries. It would already have been found. The Saudis know this, the oil companies know this, and Wall Street knows this.

Thus, if it became known that Saudi Arabia has entered decline it would rock the kingdom's standing and signal to the whole world that humankind is in deep trouble. Within the Saudi kingdom, the repercussions would be ominous among a very poor population that is largely pacified with government subsidies that are already shrinking rapidly as the oil price has been in a downward spiral from late 2008 on. The so-called recovery has brought prices back up since the spring 0f 2009 but there is little flexibility left. What the Saudis desperately need is a return to

$100 a barrel or more. But is that what the American worker needs? We will come back to this later.

There are two kinds of reserve estimates that have helped to drive us deeper into the current crisis. For purposes of this policy paper we will label them "Discovery Estimates" and "Field Reserve Estimates." Discovery Estimates are those numbers announced when a new oil field is discovered, after the first series of test wells is drilled.

Recent examples of how misleading these announcements of recent "major discoveries" show how corporations and corporate-owned media have been misleading the public.

In April of 2008 the mainstream media from Yahoo to CNN announced that based upon one test well the Carioca field in the deepwater Santos Basin off Brazil's coast, there might be 33 billion barrels of oil in the field. There were cheers, and stocks rallied as oil prices plummeted. For a minute. Nobody talked about the fact that this great news was total nonsense. Historically, when one producing well is found in a new field, a series of "appraisal" wells are always drilled around it to see how far the field might extend. As of this writing there have been no reports of any appraisal-well results. I'm certain that they are or will be drilled, but how can one calculate approximate field size and make responsible announcements without them?

To complicate matters, Carioca is a deep-sea find and deep-sea platforms and wells are very expensive, sometimes $150 million or more each, to drill. The second, third, and fourth, etc. wells haven't come in yet.

For days, all I could do was bang my head against the desk as my gullible friends told me that this was proof that Peak Oil wasn't real. They somehow didn't notice that even Brazil itself had denied the outrageous claims. The Chinese, however, did notice.

In an April 14 story entitled "Brazil's Petrobras denies giant oil field discovery," the Chinese news agency Xinhuanet wrote:

RIO DE JANEIRO, April 14 (Xinhua)—Brazil's state-owned oil company Petrobras denied Monday an earlier announcement of the discovery of a gigantic oil and gas field in southeastern Brazil.

The salt layer of the second well drilled in block BMS-9 of the announced oil field has not even been reached yet, and the huge field, if it does exist, lies below the salt layer, the company said in a statement.

The announcement of the discovery had been made earlier Monday by the director of the government's National Oil and Gas Agency Haroldo Lima. The agency is in charge of regulating the oil and gas sector in the country.

The oil field in the Santos Basin in southeastern Brazil appears to be the world's third-largest oil and gas reserve, bearing an estimated volume of 33 billion barrels, Lima said.

Petrobras' statement said that the first well drilled in the area in September 2007 has produced promising results, which have been already released to the market and still need to be confirmed by further drillings.

The drilling of the second well started on March 22 and has not yet reached the necessary depth to reach the salt layer that lies above the reserve. The layer is two km wide, according to the statement.

'The exploitation activity includes the drilling of new wells, long-lasting proofs and new geological studies to ensure the broadness of the discovery, at the end of which the results will be informed to the market,' the statement added.

The Securities and Exchange Commission of Brazil, which supervises the operation of the stock market in the country, criticized the announcement by the ANP director, which prompted a sudden climb of Petrobras' stocks on the Sao Paulo Stock Exchange (Bovespa).

> The release of relevant information made by "outsiders"
> is "harmful" to the market's operation, the commission said.

I never saw American press broadcast the correction as much as they did the discovery. Hey, let's go invest some more with Bernie Madoff!

The same pattern has been happening for years now. On Sept. 6, 2006, one successful test well in the Gulf of Mexico caused the Associated Press to herald a discovery that could "boost the nation's reserves by more than 50%." An International Herald Tribune story the same day said:

> Chevron, Devon Energy and Statoil, the Norwegian oil giant, said Tuesday that they had found 3 billion to 15 billion barrels in several fields 175 miles, or 282 kilometers, offshore.
>
> They said the oil was 30,000 feet, or 9,144 meters, below the gulf's surface, among formations of rock and salt hundreds of feet thick.
>
> While it is too early to know exactly how big the fields are, the oil companies expressed hope that they might exceed those at Prudhoe Bay, off the northern coast of Alaska.

"Hope?" One cannot fill a gas tank with hope. Let me have 20 gallons of hope please.

A BBC headline the same day read "'Huge Oil Find' in the Gulf of Mexico."

All of these stories were full of the words "could" and "might."

Years after the Gulf of Mexico stories and four months after the Brazil story, if one talks to almost any American about Peak Oil, all they can remember is these two gigantic finds and assert that there is no problem. These people are still in denial. Yet there have been no follow-up stories confirming the size of these finds or the results from any other wells that have been brought in since.

In the meantime Mexico has become an unstable, failing state as rickety as Pakistan. The drug violence being blamed for all of this ignores the fact that Mexico's oil revenues are plummeting. That's the real trigger for what is happening there.

An American president cannot make policy based on "evidence" like this which serves only to boost share prices, encourage consumption, and keep citizens in the dark. Apparently China has more respect for its people's intelligence than the United States does.

Don't trust the U.S. government either.

In 2002 my newsletter, *From The Wilderness*, published an investigation by geologist Dale Allen Pfeiffer showing that the Energy Information Administration cooked its own books (lied) to assuage any public or market concerns about oil supply. I published an excerpt from that story in my 2004 book *Crossing the Rubicon: The Decline of the American Empire at the End of the Age of Oil*:

> This [book cooking] is one of the major causes of disin-formation regarding energy issues. The US government relies on the EIA for all of its energy information. Yet the EIA, a division of the Department of Energy, has admit-ted that it reverse-engineers its studies. "These adjust-ments to the USGS [US Geological Survey] and MMS [Minerals Management Service] estimates are based on *non-technical* [emphasis mine] considerations that support domestic supply growth to the levels necessary to meet projected demand levels," stated the EIA in a report titled "Annual Energy Outlook 1998 with Projections to 2020." This means that the EIA first looks at projected figures for demand, then juggles reserve and production figures to meet that demand!
>
> Likewise, USGS reports can no longer be trusted either since the agency's about face in 2000. Prior to 2000, the USGS was talking about oil depletion and the cross-over

event between demand and supply. In 2000, however, the agency published a rosy report stating there would be abundant oil for many decades. Geologists working for the USGS have stated off the record that they do not trust USGS oil data.[3]

And the band played on. . . .

CHAPTER FIVE

Infrastructure and the Grid

There are two basic kinds of infrastructure that every nation must be concerned with. The first is energy infrastructure which is both a national and a global issue for America. Historically, a substantial part of the global investment in oil and gas that took place over seven or eight decades came from either U.S.-based companies or the U.S. government itself. The second kind of infrastructure is everything that keeps things running inside the United States: bridges, dams, sewers, roads, water supplies, and perhaps most important of all, the electric grid. Some of this domestic load is carried by the federal government, but the majority of it has traditionally been carried by state and local governments, or by public utility companies which are already breaking down (or up) for many reasons.

Take away five million or five billion dollars to buy more expensive energy for everything from school buses, garbage trucks, police cars, and ambulances to power generation and there is that much less to repair roads, sewers, and everything else we rely on, especially the grid. Take away more billions from shrinking property tax revenues and failing banks and the problems start compounding as the squeeze hits from both ends.

Our state and local governments are so broke that they have begun selling off public assets like toll roads and bridges because they can't afford to maintain them anymore. The ongoing economic collapse hasn't changed a thing. Now there is less money because tax bases are collapsing. Many state and local governments are—in the face of a just-beginning collapse—starting a sequential shut-down process.

Oil And Gas Infrastructure

On April 21, 2008, Nobu Tanaka, executive director of the International Energy Agency, gave the keynote address to the 11th International Energy Forum in Rome. In that speech he said:

> Investment is one of the main challenges we are facing in the global energy sector . . . USD 22 trillion in investment will be needed in energy-supply infrastructure by 2030. The oil sector alone needs USD 5.4 trillion. Although spending has recently increased, supply growth could remain sluggish, because of increasing costs and a proliferation of above-ground risks, such as more frequent access limitations and tighter fiscal and regulatory regimes.[1]

Twenty-two trillion (with a "T") dollars? Where is that money going to come from? The printing presses are running out of ink and ink is expensive. Do we just keep printing it? Money is needed not for investment in renewable energies and new technologies; it is needed to keep an aging oil and gas infrastructure status quo operational; to rebuild or fix pipelines, refineries, and rusty drilling rigs. It is needed to construct new offshore rigs that must be built before stopgap, unproven, and limited deepwater oil fields can be found and tapped. Remember that drilling a hole, which can be very expensive (especially in deepwater fields in the Gulf of Mexico or the Arctic), does not guarantee that oil will be there.

Investment is needed to rebuild damaged or neglected upstream oil and gas infrastructure, most critically in Iraq, which has been devastated by two major wars, a decade of UN sanctions, and internal sabotage and where the second-largest oil reserves on the planet are still on the upside of Peak.

Deepwater drilling rigs are important to the United States as it debates whether to allow new offshore drilling in formerly prohibited areas or to tap accessible fields and leases (that might hold oil) in the Gulf of Mexico and other regions. These rigs are very expensive and, as of the summer of 2008, there were virtually none available. The economic crash has limited the amount of credit and financing necessary to underwrite capital investment costs for rig building, which can only be recovered if there's oil flowing at the end of the day. The rigs themselves take years to build before drilling can even begin. The ones that have already been built are in use and—in what will come as a surprise to most—a large portion of them have already been leased by Saudi Arabia. That's right, Saudi Arabia.

The *Houston Chronicle* reported on May 2, 2006 that "by year-end, Saudi Arabia will have 120 [offshore] rigs operating in the country, up from 85 last year and 54 in 2004."[2] That is a 120% increase in offshore drilling by the Saudis in just two years.

This begs the question as to why, if Saudi Arabia insists that their onshore reserves are adequate to meet demand growth for twenty years, they are frantically exploring offshore for oil that would cost between three and ten times as much to produce as oil from their allegedly abundant onshore fields.

AMERICAN INFRASTRUCTURE

The Grid

Electricity is the foundation of modern industrial civilization. It has become so commonplace in most of the world that it is taken for granted or barely even noticed. Yet it is electricity that powers assembly lines, provides lights and air conditioning, refrigeration, telephones, computers, and TVs. Electric pumps provide all of the irrigation for our food supply as well as the water that comes out of our taps. It also pumps water out of New York City's subway system 24/7 to allow trains to operate. Electricity is what pumps gasoline into your gas tank.

The American grid has been in trouble for a long time as a result of two problems: lack of infrastructure maintenance and repair and shortages of natural gas to power generating stations which are still being built to meet new demand. Coal does provide a significant and growing portion of our electricity, but coal has brought with it the significant problem of greenhouse-gas emissions and very toxic waste, which so-called "clean coal" never acknowledges. (The term *clean coal* is a marketing gimmick because the technology does not remove the poisons from either the mining or the combustion—only the exhaust gasses. It has never been implemented commercially. I repeat . . . never in the process of commercial power generation has any so-called clean coal plant produced 1 kWh of electricity.)

There is also the problem that coal is not found everywhere in the nation, and it must be transported to those plants that use it—by rail on a railway system that has been neglected for decades. If we are going to use more coal, then we will need more trains and to fix the ones we have. More aggressive mining, including mountain-top removal, causes massive ecological damage and destroys forests. (I will have a lot more to say about coal in subsequent chapters.)

In August of 2005 President Bush signed a bill which repealed the Public Utility Holding Companies Act, or PUHCA. PUHCA was a New Deal measure that focused solely on public need. It mandated that owners of public utilities were prohibited from achieving a monopoly and—most importantly—that utility companies had to maintain excess generating capacity and infrastructure to provide for ten, twenty, and hundred-year weather events.

With the repeal of PUHCA it became possible for private investors, like Warren Buffet or Constellation Energy, to start gobbling up once publicly owned and/or regulated power companies. Just a couple of years ago Buffet gobbled up Constellation. The repeal (or deregulation) relieved the new owners of the mandate to keep excess capacity to prepare for weather-related emergencies. In the age of climate change where heat waves, floods, and severe cold snaps have become more frequent—even regular occurrences—this is just plain stupid. Also, since the repeal of PUHCA, privately owned utility companies may now selectively decide which customers get electricity and which don't. Computer technology now allows companies to decide to provide power to a major corporation across town and deny it to a rest home a block away. With deregulation, profit became not just the primary, but the sole criterion for success.

It is clear that this is what the new owners of private utilities intend. *U.S. News and World Report* wrote the following just four months after PUHCA was repealed:

> The second threat is a severe electricity shortage in the Northeast—with possible brownouts or blackouts. Deregulated natural-gas-fired power generators, *under no legal obligation to serve customers as the old monopoly electric companies were*, can simply stop generating power. Some plants will be interruptible customers with no backup fuel source. But in other cases, power plants that have firm natural gas contracts will stop generating electricity anyway and sell their fuel at enormous profit. That is precisely what happened during the three-day January 2004 cold snap, when more than 25 percent of New England's generating capacity went off line and the reserve margin was near zero [emphasis mine].[3]

Privately owned utilities are responsible to shareholders, not ratepayers. They have a completely different set of priorities and their first and governing responsibilities are profit, growth, and shareholder return. Privately owned utilities can now trade their energy reserves, whether natural gas, coal, or heating oil to other regions; not based upon need, not giving a whit about who freezes to death or whose small business is shut down. They will trade their energy for profit and for profit only. It is now legal for a privately owned Arizona power company to sell off natural gas to another power company in, say, Colorado or California just as a killer heat wave strikes the Southwest. Under a profit mandate this scenario is almost certain to occur, and people will die as a result.

There is also a trend towards the construction of "merchant power plants," smaller generating stations dedicated to serving only corporate customers. In a January 2006 essay for *From The Wilderness* called "The End of the Grid" I wrote:

> The term "merchant power plants" has come up in several stories. It suggests, though I have not been able to confirm it yet, that power companies will now be operating dedicated generating stations for industrial and corporate users with the best ability to pay. Weaker corporations, not on the "A" list, would be allowed to die-off leaving more energy for the rest. That would mean that a Boeing plant might have plenty of power sitting right next to a neighborhood that gets none at all due to selective service interruptions designed to "curb demand." As if any residential user would voluntarily have their heat and power shut off during a cold winter.[4]

Just after the great Northeast blackout in the Fall of 2003 I interviewed Matthew Simmons for *From The Wilderness*. His clients include entities like Kerr McGee and the World Bank. This is a lengthy quote but well worth reading because it addresses many of the issues we have just discussed.

> **FTW:** What did happen?
>
> **SIMMONS:** On a large scale what happened was deregulation. Deregulation destroyed excess capacity. Under deregulation, excess capacity was labeled as "massive glut" and removed from the system to cut costs and increase profits. Experience has taught us that weather is the chief culprit in events like this. The system needs to be designed for a 100-year cyclical event of peak demand. If you don't prepare for this, you are asking for a massive blackout. New

plants generally aren't built unless they are mandated, and free markets don't make investments that give one percent returns. There was also no investment in new transmission lines.

Underlying all this is the fact that we have no idea how to store electricity. And every aspect of carrying capacity, from generators, to transmission lines, to the lines to and inside your house, has a rated capacity of x. When you exceed x, the lines melt. That's why we have fuse boxes and why power grids shut down. So we have now created a vicious cyclicality that progresses over time.

Another problem was that with deregulation, people thought that they could borrow from their neighbor. New York thought it could borrow from Vermont. Ohio thought that it could borrow from Michigan, etc. That works, but only up to the point where everyone needs to borrow at once and there's no place to go.

A second major reason is that decisions were made in the 1990s that all new generating plants were to be gas-fired. We've had a natural gas summit this year and, as you know, I have been talking for some time about the natural gas cliff we are experiencing. Many thought that this winter would be deadly, and I have to say that it's just a miracle that we have replenished our gas stocks going into the cold months. This winter could have been a major disaster. We've seen a price collapse in natural gas to the five to eight dollar range (per thousand cubic feet) and the only reason that happened was throughout almost the entire summer there were only a handful of days when the temperature rose above eighty degrees anywhere. That was miraculous. It allowed us to prepare for the winter but we shouldn't be optimistic. One good hurricane that disrupts production, one blazing heat wave, one freezing winter after that and we're out of solutions.

[Note: Simmons said this two years before Hurricanes Katrina and Rita.]

FTW: And natural gas too?

SIMMONS: Well, I know you understand it, but people need to understand the concept of peaking and irreversible decline. It's a sharper issue with gas, which doesn't follow a bell curve but tends to fall off a cliff. There will always be oil and gas in the ground, even a million years from now. The question is, will you be a microbe to go down and eat the oil in small pockets at depths no one can afford or is able to drill to? Will you spend hundreds of thousands to drill a gas well that will run dry in a few months? All the big deposits have been found and exploited. There aren't going to be any dramatic new discoveries and the discovery trends have made this abundantly clear.

We are now in a box we should never have gotten into and it has very serious implications. We also see the inevitable issues that follow a major blackout: no water, no sewage, no gasoline. The gasoline issue is very important. Our gasoline stocks are at near all time lows. With the blackout, more than seven hundred thousand barrels per day of refinery capacity were shut down. People were told to boil their water. So what do they do, they go to their electric stove which isn't working. What then?[5]

Utility companies are what financial analysts call "cash cows." They produce enormous, predictable, and steady streams of cash as ratepayers pay their bills. This cash has enormous importance for private companies that it doesn't have for publicly owned companies. Because the economic paradigm calls for infinite growth, cash is used to grow private companies, whether through leveraged buy-outs of other utilities or to pay investors. Liquidity is what allows Wall Street to do "debt-service" or, in layman's terms, to make the minimum monthly payments on their credit cards. That's why utilities are so attractive for people like Warren Buffet and major investment banks.

One thing is certain. The cash generated by power companies won't go back into fixing the infrastructure, building energy reserves, or preparing for weather-related emergencies.

Roads To Ruin

On July 31, 2008, a Reuters news story proclaimed: "A $1.6 trillion bill is coming due across the United States as governments face the daunting task of repairing roads, bridges and other parts of an aging infrastructure." On the one-year anniversary of the famous I-35 bridge collapse in Minneapolis the story also noted heavy infrastructure damage as a result of 2008's heavy floods along the Mississippi River. It later added:

> State transportation officials issued an estimate this week that at least $140 billion was needed to make major repairs or upgrades to 152,000 of the nation's 590,000 bridges— one in four—deemed deficient. The spans that were built to last 50 years are on average 43 years old.[6]

Much of the funding for these needed repairs will fall on the federal government. But remember that tax revenues are plummeting at every level of government due to economic recession and shrinking tax bases.

The very next day Reuters published another story, verifying a trend I had been predicting and writing about for four years. Its title: "Roads, airports on the block as budgets tighten." This was the lead:

> NEW YORK (Reuters)—Cash-strapped US state and city governments are likely to sell or lease more highways, bridges, airports and other assets to investors desperate for stable returns after being frazzled by the credit crisis.

The trend is set to pick up speed given worsening budget deficits in state capitals and city halls nationwide.

It will also be welcomed by Wall Street bankers hoping to help create and market so-called "infrastructure" trans-actions at a time many debt markets remain paralyzed, and after major US stock indexes fell into bear market territory.

"When you are nervous about everything else, you put your money in a toll road, . . ."[7]

The Pennsylvania Turnpike is up for sale or lease (whichever is better). So is Chicago's Midway airport. New York's Governor David Patterson is looking at many options along these lines. Hey, let's sell the Washington Monument or the Statue of Liberty. There's a way to raise cash!

A close read of the story revealed that (among others) Goldman Sachs, Citigroup, Morgan Stanley, The Carlyle Group, Credit Suisse, and General Electric have already dedicated more than $25 billion to these purchases or leases of public property. Only Credit Suisse has escaped blistering criticism from me in the past on other issues. Goldman Sachs is the scariest; they seem to be the recruiting pond from which the United States draws its treasury secretaries lately. Robert Rubin and Henry Paulson both came from Goldman Sachs. Goldman started a $6.5 billion infrastructure fund in 2006 and, according to Reuters, is starting another $7.5 billion fund.

If I were a corporation, I would be tickled to death to have taxpayer money build a series of very expensive roads which my corporation or bank could then buy for pennies on the dollar (on easy credit before September 2008) and charge people fees to use. Then, when my company had run the road or airport down to the point of failure, I could always go to the federal government and ask for a bailout. Heck, everybody else is getting or asking for a bailout, even Larry Flynt!

However, by late 2009 the double-edge sword of monetary collapse had put a damper on these proposed "solutions". A

USA TODAY story on October 27th reported that "Privately run infrastructure deals dry up" as cash shortages had made it impossible for investment banks like Goldman Sachs to leverage enough capital to make the purchases. Among the deals that had vanished recently was Chicago's plan to sell Midway airport.[8]

Peak Traffic And Upside-Down Thinking

Even as private and government experts are saying clearly that cheap gasoline is transitory and a thing of the past; and while unemployed Americans are driving much less because they can't afford gasoline . . . at any price, budget experts are planning for new road expansions as if traffic were going to continue to increase as it has for the last seven decades. The United States is currently building roads almost as fast as it did in the 1950s when the Interstate Highway System was implemented as a matter of national security. Here's an example:

> MarketWatch Aug 1, 2008—By the year 2032, the U.S. population is expected to reach 363.5 million persons, adding an estimated 49 million drivers and 58 million vehicles to America's highways. Wasted fuel from traffic delays will more than double, to 6.5 billion gallons. Carbon dioxide emissions traced to congestion will increase to 60 million tons.

But there's a major rub here. U.S. traffic peaked in 2005, the same year that experts said that oil production apparently did. Americans are actually driving less, and that is a certainty where millions are unemployed and have no jobs to drive to or money to buy gasoline with, even at $2–3 a gallon. I predict that these demand-destruction, cheap prices will be a permanent thing of the past by the end of 2009 and that the next price spike in oil will serve as the coup de grace for the U.S. and world economies.

Data collected by the Bureau of Transportation statistics show that vehicle miles travelled peaked and leveled off at the onset of $3 gasoline in 2005. $4 gasoline in July 2008 caused them to markedly drop.[9] At a time when the government and every recognized expert is telling us not to expect cheap gasoline to last, and when all the oil production and depletion data suggest that we may one day look back on four-dollar gas with longing, why has the United States embarked on one of the most massive road building campaigns in history? This includes the construction of massive NAFTA Superhighways so that tomatoes can be trucked from Mexico to Canada and Canadian wood products can be driven to Mexico.

A spike, even back to $3.25 gasoline has caused as much hardship in 2009 as $4 gasoline did in 2008. And for those who hope that an economic recovery will absorb all of this, remember that no economic recovery is possible even close to where we were in early 2008 without driving oil consumption back up to where it was then. Contrast that with a 9% decline rate. The numbers just don't balance. They never have, and this is what M. King Hubbert so clearly understood in the late 1940s.

[Excellent research on the topic of unnecessary road building and airport expansion has been done by Mark Robinowitz. He maintains an excellent web site at http://www.road-scholar.org.]

With food shortages already occurring, what kind of sense does it make to pave over land that may be needed for food with asphalt made from oil? Is there some pork-barrel spending hidden away here? It sure smells like it.

Federal, state, and local governments are suffering from severe asphalt shortages caused by oil prices and demand. Streets are going unrepaired, potholes are damaging cars, and existing roads are wearing out. Why build new roads, and take on the responsibility for maintaining them, for what is certain to be fewer cars travelling fewer miles when we aren't able to repair the roads we already have?

In early 2009 one might have argued that oil's falling to around $40 would take care of the asphalt prices. But that didn't last, did it? Now go look at the budgets of state and local governments and see if there is any money left to buy even cheap asphalt. Energy and money are indeed Siamese twins.

The same holds true for many of the dozen or so major airport expansions that are being planned around the country. Why plan to expand airports when airlines are struggling to stay flying? They are cutting back flights to avoid bankruptcy, and the industry is widely predicted to have a bigger shakeout that will further reduce the number of carriers and flights.

The key to understanding infrastructure lies at the heart of complex civilizations. When roads and bridges fail; levees aren't rebuilt; when dams, transmission lines and generating stations are not maintained; when any of a hundred possible things fail for lack of money or material . . . civilization starts to break down.

Consider the implications if a bridge washes out and that bridge is the only way to get heating oil to a small city in winter. Consider the implications if a gasoline or diesel tanker truck crashes and burns on a defective bridge. Consider the implications if a main sewer line collapses in New York City and takes three subway lines out of service. Consider the implications if the electricity, generated by oil or natural gas, stops providing the power to pump water out of those subways 24/7.

The Alternative Energy Infrastructure . . .

. . . does not exist.

On top of all the maintenance needed just to keep everything running as it is, the proponents of alternative energy sources like solar, tidal and wind power almost never mention the infrastructure to make such propositions work. Plans to dramatically expand the use of wind power have yet to fully explain how thousands of miles of transmission wires and transformers are going to be built and who is going to pay for them. The world is experiencing huge commodity shortages, and copper is one of the most precious. Wind doesn't blow 24 hours a day, and the grid has no ability to store electricity generated at midnight that might be needed at 3 P.M. the next day when temperatures soar. The battery technology to store electricity on a large scale doesn't exist. Batteries are very expensive and don't last that long. Even the batteries in vaunted hybrid cars need to be replaced after 70 or 80 thousand miles.

Hydrogen remains the cruelest hoax ever perpetrated on an unsuspecting public. Hydrogen is the smallest atom in the universe: one proton and one electron. It bleeds through many metals and has a tendency to turn them brittle to the breaking point. It cannot be pumped through the hundreds of thousands of miles of existing natural gas pipelines. Those are needed for natural gas, and the pipes can't handle hydrogen. So California's Hydrogen Highway, touted by Governor Schwarzenegger, is a true pipe dream.

Solar poses similar problems. It takes large quantities of energy and resources to make solar cells. As of this writing my best research indicates that it takes about five years to get a positive return on the energy invested to manufacture the best solar panels. Solar does have one distinct and overwhelming advantage over all other alternative energies: It can be installed where it is used so there is no need for major infrastructure investments to make it work over distance. It is scalable. But what do we do for electricity at night or on a cloudy day?

The old system must be kept working.

These are questions that an American president and congress will play decisive roles in answering. As the sign on Harry Truman's desk said, "The buck stops here."

In one of my favorite all time essays, "GlobalCorp" from March of 2005, I wrote the following and it gives me no pleasure to see it coming true:

> As the human race blows itself into extinction or destroys the climate or starves itself to death, the last corporate merger and acquisition will take place. And at the same moment as mankind dies, the CFO of "GlobalCorp" will be shouting, "Hooray! We did it!"[10]

CHAPTER SIX

Iraq

This short chapter will deal with Iraq as it is; not as we might like it to be. There will never be a return to the status quo ante 2003. And there will, for the foreseeable future, always be an American military presence there and in much greater numbers than the American people might like or anticipate.

An American president is not completely free to undo the actions of previous administrations; certainly not on a unilateral basis where treaties have been enacted and laws passed by congress. The United States has already committed to being in the country with the world's second-largest known oil reserves for decades. Most importantly, Iraq, after two decades of war and sanctions—after the wholesale destruction of its oil and gas infrastructure, by neglect, war and internal sabotage—has yet to peak in oil production like most other producing nations. That peak has been delayed. Iraqi oil production reached an all-time high just before the first Gulf War in 1990 at 3.5 million barrels per day. It fell to around 300,000 barrels per day in 1992. It was back up to around two million barrels just prior to the U.S. invasion in 2003. It fell dramatically thereafter but by the summer of 2008 it had returned to around 2.5 million barrels per day.

Iraqi reserves are estimated to be around 110 billion barrels. (Again, remember that the planet is consuming a billion barrels every 11.5 days.) I prefer to use more conservative (historically accurate) estimates of around 90 billion barrels of recoverable oil which have been suggested by experienced exploration geologists, rather than market analysts. Wild projections of another 100–200 billion barrels of "undiscovered" oil in the Sunni western regions of the country don't fly much farther

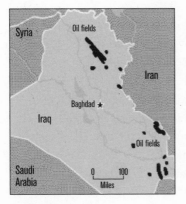

SOURCE: *USA TODAY*, JUNE 6, 2008.

than a brick with paper wings. The notion that that much oil remains undiscovered after nine decades of British, American, Iraqi, and French exploration and analysis has little traction, especially based on the always wildly overstated reserve estimates for new fields all over the world, from the Caspian, to the Gulf of Mexico, to Brazil. Even the Sunnis in whose territory the alleged oil exists know it isn't there. They fought hard to get a share of the revenues from Iraq's Kurdish North and Shiite Southeast.

New agreements negotiated between the U.S. and Iraqi governments all are based upon the reality that Iraqi oil lies in only two small regions outside of Sunni control. Saddam Hussein himself was Sunni and he would have had a much easier time developing those regions if oil had been there. Much of the "sectarian strife" that occurred between 2003 and 2008 was actually an argument about Sunnis not getting cut out of oil revenues which are now filling Iraqi "national" coffers. And, as we shall see, new agreements are in place, which include the Sunnis in Iraqi revenue sharing despite the fact that they have no oil.

The Bush-Cheney Agenda—Missions Accomplished

Consider the following:

- A *Washington Post* story in May of 2005 by Bradley Graham reported that the U.S. was building four major military bases around airfields in strategic locations and that these bases had "a more permanent character." These bases are large—very large—and heavily fortified. In addition to U.S. Air Force fighter, bomber, helicopter, and support

aircraft these fortified bases will also be home to brigade-
sized combat units plus Army Air Support and all logis-
tics and support personnel (including hospitals) to sustain
their operations.[1]

- The U.S. is nearing completion of a 21-building, fortress-
like embassy compound covering 104 acres in Baghdad at
a cost that will exceed $1 billion.[2] It is the biggest embassy
compound ever built by any nation anywhere and is larger
than Vatican City. I estimate that a regimental-sized
unit of U.S. Marines (including support which may be
outsourced) will be necessary to protect the embassy alone.

Not counting military personnel at U.S. bases in Kuwait, Qatar,
and in and around Saudi Arabia; as well as naval deployments in
the Gulf around Iraq, the total number of military personnel in
Iraq will likely never drop below 50,000. No president will be
able to or want to change this.

What was "gained" during the Bush-Cheney years will not be
given back. It is and always was about Iraqi oil; conventional oil,
much of it light-sweet crude and easy to get to after many billions
have been reinvested in Iraq's oil and gas infrastructure. That
revenue will likely come from U.S. and European oil companies,
not just Iraqi oil profits. Many have wondered what the major
U.S. oil companies are doing with their profits. And this will give
part of the answer.

Oil companies are certainly buying back shares of their own
stock to boost share prices (and eventually shut down), but they
are also awaiting the finalization of an Iraqi oil law that will allow
them to pour that money into rebuilding its infrastructure so that
future production can be raised to perhaps five or six million
barrels per day. That will never offset decline

The Bush-Cheney administration did accomplish its mission
which was threefold. First, it had to occupy the country and
make sure that neither Russia, Iran, nor any hostile regime from
within the country gained control over it.

The second objective was to partition or Balkanize the country in a way that made it less expensive to control/protect only the areas that had oil. Iraq's oil runs in a thin south-southeast to north-northwest sliver from Shiite Basra in the south to Mosul and Kirkuk in the Kurdish north. Why pay to occupy an entire country when you can break it up and only protect those areas where the oil is? The current map of Iraq was drawn in the early 1920s by Winston Churchill and other European interests with a pencil for political (oil) reasons. Although political maps will continue to show the same historic borders that Churchill drew, for all intents and purposes Iraq no longer exists as a single country. I and many others had been predicting this since at least 2002 as it became obvious the United States was going to invade no matter what.

Peter W. Galbraith confirmed this reality in an October 2007 Op-ed for the *New York Times*. He wrote:

> In a surge of realism, the Senate has voted 75–23 to acknowledge that Iraq has broken up and cannot be put back together. The measure, co-sponsored by Joe Biden, a Democratic presidential candidate, and Sam Brownback, Republican of Kansas, supports a plan for Iraq to become a loose confederation of three regions—a Kurdish area in the north, a Shiite region in the south and a Sunni enclave in the center—with the national government in Baghdad having few powers other than to manage the equitable distribution of oil revenues.
>
> While the nonbinding measure provoked strong reactions in Iraq and from the Bush administration, it actually called for exactly what Iraq's Constitution already provides—and what is irrevocably becoming the reality on the ground.
>
> The Kurdish-dominated provinces in the north are recognized in the Constitution as an existing federal region, while other parts of Iraq can also opt to form their own regions. Iraq's regions are allowed their own

Parliament and president, *and may establish their own army. (Kurdistan's army, the peshmerga, is nearly as large as the national army and far more capable.)* While the central government has exclusive control over the national army and foreign affairs, regional law is superior to national law on almost everything else. *The central government cannot even impose a tax* (emphasis mine).

Iraq's minimalist Constitution is a reflection of a country without a common identity. The Shiites believe their majority entitles them to rule, and a vast majority of them support religious parties that would define Iraq as a Shiite state. Iraq's Sunni Arabs cannot accept their country being defined by a rival branch of Islam and ruled by parties they see as aligned with Iran. And the Kurdish vision of Iraq is of a country that does not include them . . .

So we should stop arguing over whether we want "partition" or "federalism" and start thinking about how we can mitigate the consequences of Iraq's unavoidable breakup. Referendums will need to be held, as required by Iraq's Constitution, to determine the final borders of the three regions. There has to be a deal on sharing oil money that satisfies Shiites and Kurds but also guarantees the Sunnis a revenue stream, at least until the untapped oil resources of Sunni areas are developed. And of course a formula must be found to share or divide Baghdad.[3]

Note that there are still significant details to be worked out. This is a major reason why U.S. troops will have to stay; to keep remaining piles of kindling from igniting and, most importantly, to make sure that other powers (e.g. Russia and Iran) don't gain significant influence.

The U.S. military, under the command of George W. Bush, had been thinking this way for some time also. The following map was prepared by the *Armed Forces Journal* in June 2006 and posted on its web site:

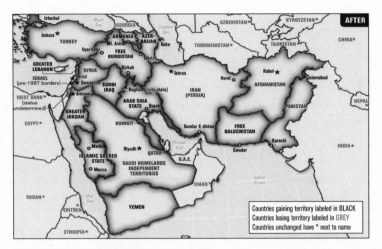

SOURCE: RALPH PETERS, *ARMED FORCES JOURNAL*.

Although the map has since been removed it was quickly copied and reposted on other web sites for posterity because it so accurately reflected predictions many of us had been making for close to four years. The story that went with it was titled "Blood Borders—How a Better Middle East Would Look."[4]

From as far back as 2003 this agenda was made clear by Council on Foreign Relations member Leslie Gelb in a *New York Times* Op-ed entitled "The Three State Solution":

> President Bush's new strategy of transferring power quickly to Iraqis, and his critics' alternatives, share a fundamental flaw: all commit the United States to a unified Iraq, artificially and fatefully made whole from three distinct ethnic and sectarian communities. That has been possible in the past only by the application of overwhelming and brutal force.
>
> President Bush wants to hold Iraq together by conducting democratic elections countrywide. But by his daily reassurances to the contrary, he only fans devastating rumors of an American pullout. Meanwhile, influential senators have called for more and better American troops to defeat the

insurgency. Yet neither the White House nor Congress is likely to approve sending more troops.

And then there is the plea, mostly from outside the United States government, to internationalize the occupation of Iraq. The moment for multilateralism, however, may already have passed. Even the United Nations shudders at such a nightmarish responsibility.

The only viable strategy, then, may be to correct the historical defect and move in stages toward a three-state solution: Kurds in the north, Sunnis in the center and Shiites in the south. . . . [5]

The third mission of the Bush administration was to see to it that agreements were in place that allowed the United States, its oil industry, and its allies: 1) access to Iraqi oil; 2) a dominant share of the infrastructure work; and 3) control over where most of the oil would be sold and to whom. Those agreements are already in place; though relatively little is known about them.

Not all of that oil, so essential to the entire planet as depletion accelerates, will go to the United States. As production declines around the world, industrialized and industrializing nations will need their share. As the United States will control the oil, this will (as intended) provide a powerful tool for the United States to bond allies closer and punish or contain would-be adversaries. In a few short years Iraq will be the only swing-producing nation with the capability of sending out buckets to address the severed artery of global depletion.

A *Los Angeles Times* story from July 2008 said that of 35 companies bidding on contracts to rebuild Iraqi oil infrastructure "among them were seven from the United States and four each from China and Japan."[6]

China gets a share. Japan—which has zero oil reserves—gets a share to keep its economy running and to keep it from collapsing and falling under (inevitable) Chinese regional dominance. Those two countries are far and away the largest purchasers of

U.S. Treasury notes which keep the U.S. economy (barely) functioning. If their economies fail for lack of energy, there is no one
capable of financing US-debt.

The third mission of the Bush administration then has been
to make sure that US oil companies get their feet securely
planted in Iraqi oil fields. As of this writing that mission is only
partly accomplished; though, it looks well on the way. The Iraqi
Hydrocarbon Law had not yet been sent to the Iraqi parliament
(fall 2008) and only portions of it have been disclosed. Other
portions, which specify how ownership, revenue, and profits
will be allotted, appear to be in flux within the Iraqi cabinet
and its American protectors. In June of 2006 U.S. Democratic
senators Kerry and Schumer sent a letter to Secretary of State
Condoleezza Rice asking her to prevent the various autonomous
regions within Iraq from signing no-bid contracts with American
companies.[7]

One section of the hydrocarbon law that has been disclosed
says that the Sunni region will get a substantial share of all oil
revenue through the national oil ministry which is the only real
national government that still exists.

Though beyond the scope of this book to discuss in detail, it
appears that U.S. companies are confident about their Iraqi future.
We do not know what deals have already been made or how they
will be presented to the Iraqi central government and the autonomous regions. I suspect that Production Sharing Agreements
(PSAs) will be put into place that will guarantee U.S. companies
access to both oil and profits for perhaps thirty years, after which
time further negotiations will be essentially moot.

Iraqi oil production is rising in the relative calm since the U.S.
military "surge" of 2007–2008, and it is likely that robust U.S.
influence in Kurdistan will secure beneficial deals for the U.S. and
NATO countries there while deals in the Shiite regions will prove
beneficial for the United States, China, Japan, and other regions.

Any U.S. president after 2008 should strive for two things:
equitable distribution of Iraq's oil to all countries who wish to

buy it, and protection of Iraq's oil for the benefit of the various Iraqi autonomous regions in such a way as to ensure stability in the region.

The worst thing an American energy policy could do would be to take the position that the oil in Iraq belongs to the United States at bargain-basement prices. That would accomplish only a renewed war inside the country, uniting the various factions against us, and it would turn the entire world, crumbling under depletion in the face of increased demand and collapsing economies, against us. The United States has spent almost all of its political, economic and military capital to create a new status quo in the Middle East. It cannot afford to do anything else but play the cards in hand today.

It was an unjust invasion; an invasion initiated on a fabric of lies and a subsequent war of unspeakable brutality with massive and senseless civilian casualties. But no subsequent American president will have the power, or the resources, to go back and start all over again.

[Author's Note: On February 27, 2009, the Obama White House announced that instead of a permanent withdrawal, as many as 50,000 U.S. troops would remain permanently stationed in Iraq. I finished writing this chapter in July of 2008 but had actually been saying that this was the plan since the invasion in 2003. I decided to leave the chapter the way I wrote it. —MCR]

CHAPTER SEVEN

Saudi Arabia

It is undisputed that the Kingdom of Saudi Arabia contains roughly 25% of the known conventional oil reserves on the planet. There is increasing hard evidence that Saudi Arabia has passed or is about to pass its production peak and enter what might prove to be a very steep decline curve.

OPEC camouflages decline by calling it production cutbacks to boost prices.

Saudi Arabia is the most difficult challenge for U.S. foreign policy from an energy perspective. From the end of World War II it was basically a U.S. client-state, but those tables have since been reversed in many arenas, not only by virtue of American dependence upon Saudi oil but also upon Saudi investment in the U.S. economy. Saudi Arabia is currently the second-largest oil supplier to the United States, and Saudi nationals have major investments in some of the most powerful U.S. corporations and banks.

A 2003 paper by the Saudi American forum reported that 60% of all Saudi foreign investments had been placed in the United States.[1] In 2002 after the attacks of 9/11, victim families threatened to sue Saudi Arabia as the U.S. media focused on ties between the hijackers and members of the Saudi government, not to mention investors from the multi-billion-dollar Saudi "Bin Laden Group" (a construction conglomerate owned by relatives of Osama bin Laden). Suddenly the Saudis started dropping hints that they might withdraw their investments in the United States. That virtually killed further news coverage of these questions and sent shockwaves through U.S. financial markets. So much for a fearless, independent media.

As I documented from many open sources in my 2004 book *Crossing the Rubicon*:

> It is impossible to quantify exactly the Saudi holdings in the US economy. But anecdotal evidence is compelling. *The New York Times* reported on Aug. 11, "An adviser to the Saudi royal family made a telling point about Saudi elites. He said an estimated $600 billion to $700 billion in Saudi money was invested outside the kingdom, a vast majority of it in the United States or in United States-related investments." The BBC has estimated Saudi-US investment at $750 billion.
>
> Adnan Khashoggi, perhaps the best-known Saudi billionaire, controls his investments through Ultimate Holdings Ltd. and in Genesis Intermedia. . . . The rest of his private US holdings are administered . . . from offices in Tampa, Florida, not far from where many of the hijackers received flight training at both private schools and US military installations.
>
> Khashoggi is a longtime financial player deeply connected to the Iran-Contra scandal of the 1980s and also to BCCI [Bank of Credit and Commerce International]. But Khashoggi doesn't even make the Forbes list of the richest people in the world. One Saudi who does is Prince Alwaleed Bin Talal, who ranks as the eleventh richest man on the planet with an estimated net worth of $20 billion. (Alwaleed is also an investor in and reported client of the Carlyle Group.)[39]

Some of Alwaleed's holdings and recent acquisitions include:

- The single largest shareholder in Citigroup, the teetering U.S. financial giant, which is reported to have a derivatives bubble of more than $12 trillion and has reportedly sought recent emergency assistance from the Federal Reserve.

On July 18 Alwaleed made an additional $500 million purchase of Citigroup stock, raising his estimated share-holding to $10 billion. (The BCCI scandal was not the last instance where the prohibited foreign ownership of U.S. banks was an issue that touched Saudi interests. Prince Alwaleed's heavy stake in Citigroup was concealed from 1991 until recently by the Carlyle group, which, acting as a virtual cutout, disguised Alwaleed's heavy investment in the bank.)[40]

- Alwaleed also owns, according to an August 9, 2002 story in the Guardian, 3% of the total shares of Newscorp (FOX), making him the second-largest shareholder behind Rupert Murdoch.
- Alwaleed's other significant holdings include Apple Computer, Priceline, the Four Seasons Hotels, Planet Hollywood, Saks, and Euro Disney.
- Alwaleed also sits on the board of directors of the infamous Carlyle Group.[2]

The Saudis have been increasing their share of investment in the Chinese, Indian, and other economies. The Saudis need not fear a United States cessation of oil purchases. It wouldn't hurt them at all now. The biggest remaining engines of growth are now outside the United States. Even if U.S. purchases stopped tomorrow (an impossibility), every drop of Saudi oil would be quickly snapped up by other customers. According to the U.S. Department of Energy Saudi Arabia is currently the second-largest supplier of crude oil to the United States after Canada, which has recently moved up a notch in the wake of severe Mexican decline. As of July 2008 America's top five oil suppliers are (in order): Canada, Saudi Arabia, Mexico, Venezuela, Nigeria, and Iraq.[3] I expect Mexico to drop one or two places in the next year, which would move Venezuela into the number 2 position.

Saudi Arabia is an extremely unstable state. Held together by brute force and social welfare handouts, the population is under-educated, restive and increasingly influenced by radical theology which is opposed to the corruption and nepotism of the royal family. Its shrinking seas of oil await only a spark to bring the kingdom down. One of the Saudi government's (indeed the world's) greatest worries is a terrorist attack on its largest refineries and the ports of Ras Tanura and Juaymah from which all Saudi exports leave for gas tanks on five continents.

When I wrote *Crossing the Rubicon* in 2002–2004 I documented how so much of Saudi oil wealth was going to the royals that the Saudi government had actually been borrowing money to fund its social programs in an attempt to keep order. The dramatic increase in oil prices since then temporarily solved this problem for the House of Saud, but the basic fault lines remain intact and aggravated. Now collapsed oil prices threaten stability more than any event in recent history. The Saudi monarchy did not—I can guarantee you—pay down its "credit cards" when oil was over $100. Oil will inevitably approach or break through $100 a barrel if the economic recovery returns GDP to pre-crash levels either in 2009 or 2010. If "recovery" really happens (meaning GDP starts to surpass 2008 levels, the $200 or $300 oil is inevitable. But all every oil-producing nation will do then is rush to play catch up for a brief moment before demand collapses all over again.

The Saudi royal family is as worried about decline and collapse as the rest of the world. More so, they are worried about any news of that reaching its subjects when it happens. Two things are important to remember. Saudi Arabia did not exist as a nation before the early 1930s. Its borders were drawn in much the same way as Iraq's. Second, Saudi Arabia is home to the two most important religious sites in all of Islam, the world's second-largest religion: Mecca and Medina.

A Brief History Of Aramco

The company, now called Saudi Aramco, was originally a U.S. enterprise. That explains why U.S. experts like Matthew

Simmons have such a good foundation for analyzing Saudi production and field life. Here's a good summary from Wikipedia:

> Saudi Aramco's history dates back to May 29, 1933 when the Government of Saudi Arabia signed a concessionary agreement with Standard Oil of California (Socal) allowing them to explore Saudi Arabia for oil. Standard Oil of California passed this concession to a wholly-owned subsidiary called California-Arabian Standard Oil Co. (Casoc). In 1936 with the company having no success at locating oil, the Texas Oil Company (Texaco) purchased a 50% stake of the concession.
>
> After a long search for oil that lasted around four years without success, the first success came with the seventh drill site in Dammam, an area located a few miles north of Dharan in 1938, a well referred to as Dammam number 7. The discovery of this well, which immediately produced over 1,500 barrels per day (240 m³/d), gave the company the confidence to continue and flourish. The company name was changed in 1944 from California-Arabian Standard Oil Company to Arabian American Oil Company (or Aramco). In 1948 Standard Oil of California and the Texas Oil Company were joined as investors by Standard Oil of New Jersey who purchased 30% of the company, and Socony Vacuum who purchased 10% of the company, leaving Standard Oil of California and the Texas Oil Company with equal 30% shares.
>
> In 1950, King Abdul Aziz ibn Saud threatened to nationalize his country's oil facilities, thus pressuring Aramco to agree to share its profits on oil sales 50/50. A similar process had taken place with American oil companies in Venezuela few years earlier. The American government granted U.S. Aramco member companies a tax break known as the Golden gimmick equivalent to the profits lost in sharing oil profits with Ibn Saud.

In 1973 the Saudi Arabian government acquired a 25% share of Aramco, increased this to 60% by 1974 and finally acquired full control of Aramco by 1980. In November 1988 the company changed its name from Arabian American Oil Company to Saudi Arabian Oil Company (or Saudi Aramco).

The single greatest need for the entire planet is to know the true status of Saudi fields and reserves. A collapse in Saudi production would send the global and U.S. economies reeling (further) in an instant. For many reasons the Saudis have been very reluctant to share (publicly at least) what they know. This information may already be in U.S. hands, however highly classified and, therefore, beyond the reach of the public.

As I began updating years of my own research for this book I found a serious dearth of new information on Saudi oil reserves and economic activity, especially concerning Saudi investment in the United States. Since *Crossing the Rubicon* was published in 2004, few have delved seriously into the subject with one very notable exception. A book by Matthew Simmons used hard data on field by field production to make a strong case that many Saudi fields were in decline and perhaps near collapse because of overly aggressive production. Simmons' book hasn't been challenged and remains the premier desk reference book on the subject of Saudi reserves. Simmons has also become a regular guest commentator on CNN and other networks.

The United States is the single largest supplier of military aid to Saudi Arabia. It maintains massive military installations in and around the country. Indeed, the monarchy can be said to exist only because of U.S. military might. Ironically perhaps, these U.S. bases, greatly reinforced in the run-up to the Iraqi invasion, are also ideally positioned to deploy massive military might into Saudi Arabia should the need arise.

In his State-of-the-Union address in 1980, President Jimmy Carter set forth The Carter Doctrine which says essentially that the U.S. government will not hesitate to use military force to protect its interests in the Persian Gulf. All U.S. military deployment and expansion in the region since has been based on it.

If the United States needs oil from Saudi Arabia and information about when Saudi fields might enter steep decline, Saudi Arabia needs protection from the United States against enemies within and without the country. Though there seems to be little an American president can do to influence the Saudis, it is this fact which offers future leadership a little leverage. It also appears that the U.S. military has prepared to partition Saudi Arabia, just like Iraq, when it becomes convenient. Go back and take a look at the map on page 63.

I predicted this in 2003.

CHAPTER EIGHT

Food

Nature does not listen to "spin."

"Momma, why do our strawberries come from Chile, our
spinach from China, and our tomatoes from Mexico?
Can't we grow them here if oil is so expensive?"

If there is anything that must be understood with regard to
energy it is its relationship to food. Food does not materialize
out of thin air. It grows, either directly or indirectly, from the
soil with water and sunlight. All growth requires energy. The
animals we consume eat plants before we eat them. Soil is the
place where the plants we eat get their food. It is not an inexhaustible resource. If the soil is not healthy and full of nutrients,
or without water, it will grow nothing. Plants will starve. We will
have nothing to eat. America's population is vastly larger than it
was at the turn of the Twentieth Century when most Americans
were farmers and our soil was healthy and organic (i.e. natural).

Almost all of the arable land in this country is now used either
for commercial agriculture or for residential purposes. Worse,
much of our arable land has been rendered incapable of growing
food without massive inputs of oil and natural gas, the very items
we are running out of.

The soil must be fed on a regular basis, just as plants, animals,
and humans must. But with what? How do nutrients get back
into the soil after plants extract them to grow?

In several ways, but the most important is the return of plant
matter to the soil where it can decompose and return essential
compounds. This applies to what is euphemistically (and incor-

rectly) called "plant waste" by major corporations. This would include leaves, corn husks, wood chips, and the leftovers from sugar cane; anything that is not eaten. This is the essence of the common backyard compost heap, long recognized as about the best fertilizer around.

Another way that plant matter is returned to the soil is through manure. Animals eat plants, transform them, and return many of the most essential elements (including seeds) through their droppings. This is the way the nature sowed new crops for millions of years. Historically there was always a balance in that process which limited human population growth. That balance was broken when man first tilled the soil, tens of thousands of years ago.

What remained of that balance has been vaporized in the last century.

Over the last sixty years mankind has taken to artificially replenishing soil nutrients with chemicals derived from oil and natural gas. Natural gas is the feedstock for all nitrogen-based fertilizers used in commercial agriculture. It is used first to make ammonia, which is then transformed into fertilizer. It greatly increases productivity but it is not sustainable and actually harms the soil.

This "green revolution" as it was called in the 1950s and 1960s is what accelerated the massive population explosion over the last century from just over one billion to six and a half billion people. It started when oil-powered tractors, harvesters, petrochemicals, and electricity multiplied the amount of land that could be tilled, planted, irrigated, and harvested. It really took off when we found out how to turn oil and gas into "steroids" to make the soil work harder and turned it into an addict that could not function without them.

Take away the steroids and the soil is barren.

Ancient civilizations perished because they did not rotate their crops so that different plants would draw different compounds from the soil and return others back. The ancients did not know how to leave parcels of land fallow, one year out of three, four or five, to regenerate naturally by resting. Like humans, the soil needs a "day" off now and again. Modern agriculture, using oil and natural gas, has replaced that practice with monocropping: the practice of re-growing the same crop (e.g. corn) over and over again, year after year, on the same patch of land until the soil is useless for anything without petrochemicals.

If there is any issue which will reveal whether American governments are serving corporations or the people it is food. More specifically, it will be an issue of whether we support the use of food (e.g. corn) or other plant matter to make fuels to power internal combustion engines. It may be called ethanol, biodiesel, flex fuel, or something else but it is all the same thing, and it is threatening our ability to eat. Faced with a choice of eating or driving, most people would not hesitate to choose eating.

The process of turning *any* plant matter into fuel is also extremely inefficient, especially when compared to the fuels we have known. While debate persists as to the degree of inefficiency, there is little doubt that, once one takes into account the fuels and energy needed to plow, irrigate, fertilize, harvest, and convert crops into biofuels, it requires almost as much or more energy to make ethanol as one gets from burning it. The entire U.S. subsidy program for ethanol production is short-sighted and nothing but a gift to corporations and agribusiness that only reduces our ability to adapt to declining energy supplies.

Already food has become a major issue for the United States, and some of these issues are even moving to center stage. Consider the following recent news stories.

Here's a recent headline from a story in the *New York Times*: "Corn Farmers Smile as Ethanol Prices Rise, but Experts on Food Supplies Worry."[1] That simple statement opens the door for uglier realities.

Almost everyone is suffering under rising food prices. And mainstream media is reporting all over the country what is now obvious. Generous subsidies to grow corn for ethanol, along with rising prices, are hurting Americans' ability to buy food. Cows eat corn. That's where hamburgers and steak come from. Every kind of livestock eats some kind of plant. We are adding fuel to our tanks and taking food from our bellies. Given obesity rates in this country some would argue that this is a good thing. But at what cost?

Here's another more recent story from the Friday June 6, 2008, *Tri-State* Observer,* "The US Has No Remaining Grain Reserves."

> According to the May 1, 2008 CCC [Commodity Credit Corporation] inventory report, there are only 24.1 million bushels of wheat in inventory, so after this sale there will be only 5.73 million bushels of wheat left the entire CCC inventory. . . . "Our concern is not that we are using the remainder of our strategic grain reserves for humanitarian relief. AAM fully supports the action and all humanitarian food relief.
>
> "Our concern is that the U.S. has nothing else in our emergency food pantry. There is no cheese, no butter, no dry milk powder, no grains or anything else left in reserve. The only thing left in the entire CCC inventory will be 5.73 million bushels of wheat which is about enough wheat to make about ½ of a loaf of bread for each of the 300 million people in America."

The *Christian Science Monitor* ran this story by Colin A. Carter and Henry I. Miller from the May 21, 2007, edition:

HIDDEN COSTS OF CORN-BASED ETHANOL: DIVERTING CORN FROM FOOD TO FUEL COULD CREATE UNPRECEDENTED TURMOIL

> Policymakers and legislators often fail to consider the law of unintended consequences. The latest example is their

---

* Pennsylvania, New York, New Jersey

attempt to reduce the United States' dependence on imported oil by shifting a big share of the nation's largest crop—corn—to the production of ethanol for fueling automobiles.

Good goal, bad policy. In fact, ethanol will do little to reduce the large percentage of our fuel that is imported (more than 60 percent), and the ethanol policy will have ripple effects on other markets. Corn farmers and ethanol refiners are ecstatic about the ethanol boom and are enjoying the windfall of artificially enhanced demand. But it will be an expensive and dangerous experiment for the rest of us.

On Capitol Hill, the Senate is debating legislation that would further expand corn ethanol production. A 2005 law already mandates production of 7.5 billion gallons by 2012, about 5 percent of the projected gasoline use at that time. *These biofuel goals are propped up by a generous federal subsidy of 51 cents a gallon for blending ethanol into gasoline* and a tariff of 54 cents a gallon on most imported ethanol to help keep out cheap imports from Brazil (emphasis mine).

President Bush has set a target of replacing 15 percent of domestic gasoline use with biofuels (ethanol and biodiesel) during the next 10 years, which would require almost a fivefold increase in mandatory biofuel use, to about 35 billion gallons. With current technology, almost all of this biofuel would have to come from corn because there is no feasible alternative. *However, achieving the 15 percent goal would require the entire current US corn crop, which represents a whopping 40 percent of the world's corn supply.* This would do more than create mere market distortions; the irresistible pressure to divert corn from food to fuel would create unprecedented turmoil (emphasis mine).

The Earth Policy Institute perhaps said it best in this January 24, 2008, article by Lester R. Brown:

WHY ETHANOL PRODUCTION WILL DRIVE WORLD
FOOD PRICES EVEN HIGHER IN 2008

We are witnessing the beginning of one of the great trag-
edies of history. The United States, in a misguided effort
to reduce its oil insecurity by converting grain into fuel for
cars, is generating global food insecurity on a scale never
seen before.

The world is facing the most severe food price inflation
in history as grain and soybean prices climb to all-time
highs. Wheat trading on the Chicago Board of Trade on
December 17th breached the $10 per bushel level for the
first time ever. In mid-January, corn was trading over $5
per bushel, close to its historic high. And on January 11th,
soybeans traded at $13.42 per bushel, the highest price
ever recorded. All these prices are double those of a year
or two ago.

As a result, prices of food products made directly from
these commodities such as bread, pasta, and tortillas, and
those made indirectly, such as pork, poultry, beef, milk,
and eggs, are everywhere on the rise. In Mexico, corn meal
prices are up 60 percent. In Pakistan, flour prices have
doubled. China is facing rampant food price inflation, some
of the worst in decades.

In industrial countries, the higher processing and
marketing share of food costs has softened the blow, but
even so, prices of food staples are climbing. By late 2007,
the U.S. price of a loaf of whole wheat bread was 12
percent higher than a year earlier, milk was up 29 percent,
and eggs were up 36 percent. In Italy, pasta prices were up
20 percent.

Consider this from the *New York Times*, Monday, June 30, 2008:

"HOARDING NATIONS DRIVE FOOD
COSTS EVER HIGHER"

BANGKOK—At least 29 countries have sharply curbed food exports in recent months, to ensure that their own people have enough to eat, at affordable prices.

When it comes to rice, India, Vietnam, China and 11 other countries have limited or banned exports. Fifteen countries, including Pakistan and Bolivia, have capped or halted wheat exports. More than a dozen have limited corn exports. Kazakhstan has restricted exports of sunflower seeds.

The restrictions are making it harder for impoverished importing countries to afford the food they need. The export limits are forcing some of the most vulnerable people, those who rely on relief agencies, to go hungry. . . .

And by increasing perceptions of shortages, the restrictions have led to hoarding around the world, by farmers, traders and consumers. . . .

"Every country must first ensure its own food security," said Kamal Nath, the minister of commerce and industry in India, which has barred exports of vegetable oils and all but the most expensive grades of rice. . . .

Historically, the United States and Canada have been known as the world's breadbasket. Since the end of World War II our grain surpluses have been shipped all around the world on such a regular and dependable basis that entire nations—dozens of them—became dependent on these gifts. It was a great foreign policy tool during the Cold War. But in 2008, as a result of soil depletion and declining production, and our own rising populations, that relationship is ending.

The American web site Suite 101 recently published the following statistics on corn.[2]

The following countries accounted for 84% of the U.S. $11.2 billion in corn exports last year.

1. Japan . . . U.S. $ 2.6 billion (23.6% of total U.S. corn exports)
2. Mexico . . . $2.1 billion (18.7%)
3. South Korea . . . $830 million (7.4%)
4. Taiwan . . . $793 million (7.1%)
5. Egypt . . . $662.8 million (5.9%)
6. Colombia . . . $557.7 million (5%)
7. Canada . . . $494.8 million (4.4%)
8. Israel . . . $169 million (1.5%)
9. Morocco . . . $156.2 million (1.4%)
10. Turkey . . . $154.8 million (1.4%)
11. Saudi Arabia . . . $147.9 million (1.3%)
12. Chile . . . $94.9 million (0.8%)
13. Venezuela . . . $91.3 million (0.8%)
14. Ecuador . . . $91 million (0.8%)
15. Indonesia . . . $88.8 million (0.8%)
16. Peru . . . $70.9 million (0.6%)
17. Panama . . . $62.2 million (0.6%)
18. Malaysia . . . $59.9 million (0.5%)
19. Spain . . . $54.3 million (0.5%)
20. Ireland . . . $51.2 million (0.5%)

On August 13, 2008, *The Chicago Tribune* reported that 30% of the year's "bumper" crop harvest would go to produce ethanol.[3] With a 2008 "bumper crop" one is tempted to think everything is OK then. But let's go back and realize that the United States has set standards for ethanol production that will require *the entire U.S. corn crop in just a few years*, leaving Americans nothing to eat.

Are You As Smart As A Fifth Grader?

Many of these nations have been able to eat as a result of U.S. aid or sales of our subsidized crops for half a century. So then, has it become official U.S. policy that Japan, Mexico, South Korea, and all these nations should starve so that we can drive? What about the countries that depend upon U.S. food aid? Should we tell the people in famine-ridden Darfur that we can send no more food because of our need to drive?

Wait a second, we're telling our own people the same thing, because U.S. food banks to help the poor are empty too.

What about grains like wheat and oats? Gee, that's already in short supply. And the U.S. population is expected to be more than 520 million by the end of this century. The ramifications of this simple arithmetic are endless and none of them are good.

Hawaii in 2008 was an excellent case in point. In June and July stories began airing on TV stations that it costs $6 for a gallon of milk, $8 for a gallon of orange juice and $3 to $4 for a loaf of bread. Why? Hawaii has to import all of its food. The situation has been worsened by globalization. I recalled seeing a story in 2007 saying that Hawaii's last dairy had closed because it could not compete with globalized corporate dairy farming outside the islands.

Globalization will die with ever-increasing fuel costs. That is a good thing. And all regions of the world will have to resort to localized food production. But for the time being there are no employed dairy cows in Hawaii unless they are privately owned. How much energy will be required to start up a new dairy farm when it becomes necessary?

Why are we destroying food production capacity at a time when the entire human food chain needs to be rebuilt from the ground up?

Food Is Not A Fuel—Plant "Waste" Is Not A Fuel

What you just read required only basic skills and education. However, the implications of the changing relationship between food and energy signal an even deeper crisis. The following will require diligent, attentive reading at a college level. If you do not read at that level, then please stop, take a break and come back fresh with a dictionary. What is written here is worth a hundred times the energy you will invest to understand it. This is that important; to you, to your family, and to our nation.

My newsletter, *From The Wilderness*, first published the following article by geologist and scientist Dale Allen Pfeiffer in 2003. It remains the most frightening story we published in our eight and a half years. I have edited it only slightly for this book.

> [Note: Some months ago, concerned by a Paris statement made by Professor Kenneth Deffeyes of Princeton regarding his concern about the impact of Peak Oil and Gas on fertilizer production, I tasked *FTW*'s Contributing Editor for Energy, Dale Allen Pfeiffer, to start looking into what natural gas shortages would do to fertilizer production costs. His investigation led him to look at the totality of food production in the US. Because the US and Canada feed much of the world, the answers have global implications.
>
> Even as we have seen CNN, Britain's Independent and Jane's Defence Weekly admit the reality of Peak Oil and Gas within the last week, acknowledging that world oil and gas reserves are as much as 80% less than predicted, we are also seeing how little real thinking has been devoted to the host of crises certain to follow; at least in terms of publicly accessible thinking.
>
> All told, Dale Allen Pfeiffer's research and reporting confirms the worst of *FTW*'s suspicions about the consequences of Peak Oil, and it poses serious questions about what to do next. Thus far, it is clear that solutions for these questions, perhaps the most important ones facing mankind, will by necessity be found by private individuals and communities, independently of or outside governmental help. Whether the real search for answers comes now, or as the crisis becomes unavoidable, depends solely on us.
> —MCR]

EATING FOSSIL FUELS
BY DALE ALLEN PFEIFFER

October 3, 2003, 1200 PDT, (*FTW*)—Human beings (like all other animals) draw their energy from the food they eat. Until the last century, all of the food energy available on this planet was derived from the sun through photosynthesis. Either you ate plants or you ate animals that fed on plants, but the energy in your food was ultimately derived from the sun.

It would have been absurd to think that we would one day run out of sunshine. No, sunshine was an abundant, renewable resource, and the process of photosynthesis fed all life on this planet. It also set a limit on the amount of food that could be generated at any one time, and therefore placed a limit upon population growth. Solar energy has a limited rate of flow into this planet. To increase your food production, you had to increase the acreage under cultivation, and displace your competitors. There was no other way to increase the amount of energy available for food production. Human population grew by displacing everything else and appropriating more and more of the available solar energy.

The need to expand agricultural production was one of the motive causes behind most of the wars in recorded history, along with expansion of the energy base (and agricultural production is truly an essential portion of the energy base). And when Europeans could no longer expand cultivation, they began the task of conquering the world. Explorers were followed by conquistadors and traders and settlers. The declared reasons for expansion may have been trade, avarice, empire or simply curiosity, but at its base, it was all about the .expansion of agricultural productivity. Wherever explorers and conquistadors traveled, they may have carried off loot, but they left plantations. And settlers [and slaves] toiled to clear land and establish their own homestead. This conquest and expansion went on until there was no place left for further expansion. Certainly, to this day, landowners and farmers fight to claim still more land for agricultural productivity, but they are fighting over crumbs. Today, virtually all of the productive land on this planet is being exploited by agriculture. What remains unused is too steep, too wet, too dry or lacking in soil nutrients.[4]

Just when agricultural output could expand no more by increasing acreage, new innovations made possible a more thorough exploitation of the acreage already available. The process of "pest" displacement and appropriation for agriculture accelerated with the industrial revolution as the mechanization of agriculture hastened the clearing and tilling of land and augmented the amount of farmland which could be tended by one person. With every increase in food production, the human population grew apace.

At present, nearly 40% of all land-based photosynthetic capability has been appropriated by human beings.[5] In the United States we divert more than half of the energy captured by photosynthesis.[6] We have taken over all the prime real estate on this planet. The rest of nature is forced to make due with what is left. Plainly, this is one of the major factors in species extinctions and in ecosystem stress.

The Green Revolution

In the 1950s and 1960s, agriculture underwent a drastic transformation commonly referred to as the Green Revolution. The Green Revolution resulted in the industrialization of agriculture. Part of the advance resulted from new hybrid food plants, leading to more productive food crops. Between 1950 and 1984, as the Green Revolution transformed agriculture around the globe, world grain production increased by 250%.[7] That is a tremendous increase in the amount of food energy available for human consumption. This additional energy did not come from an increase in incipient sunlight, nor did it result from introducing agriculture to new vistas of land. The energy for the Green Revolution was provided by fossil fuels in the form of fertilizers (natural gas), pesticides (oil), and hydrocarbon fueled irrigation.

The Green Revolution increased the energy flow to agriculture by an average of 50 times the energy input of traditional agriculture.[8] In the most extreme cases, energy consumption by agriculture has increased 100 fold or more.[9]

In the United States, 400 gallons of oil equivalents are expended annually to feed each American (as of data provided in 1994).[10] [That was approximately 3 billion barrels of oil equivalent per year in 1994.] Agricultural energy consumption is broken down as follows:

- 31% for the manufacture of inorganic fertilizer
- 19% for the operation of field machinery
- 16% for transportation
- 13% for irrigation
- 08% for raising livestock (not including livestock feed)
- 05% for crop drying
- 05% for pesticide production
- 08% miscellaneous[8]

Energy costs for packaging, refrigeration, transportation to retail outlets, and household cooking are not considered in these figures.

To give the reader an idea of the energy intensiveness of modern agriculture, production of one kilogram of nitrogen for fertilizer requires the energy equivalent of from 1.4 to 1.8 liters of diesel fuel. This is not considering the natural gas feedstock.[11] According to The Fertilizer Institute, in the year from June 30 2001 until June 30 2002 the United States used 12,009,300 short tons of nitrogen fertilizer.[12] Using the low figure of 1.4 liters diesel equivalent per kilogram of nitrogen, this equates to the energy content of 15.3 billion liters of diesel fuel, or 96.2 million barrels.

Of course, this is only a rough comparison to aid comprehension of the energy requirements for modern agriculture.

In a very real sense, we are literally eating fossil fuels. However, due to the laws of thermodynamics, there is not a direct correspondence between energy inflow and outflow in agriculture. Along the way, there is a marked energy loss. Between 1945 and 1994, energy input to agriculture increased 4-fold while crop yields only increased 3-fold.[13] *Since then, energy input has continued to increase without a corresponding increase in crop yield. We have reached the point of marginal returns.* Yet, due to soil degradation, increased demands of pest management and increasing energy costs for irrigation (all of which is examined below), modern agriculture must continue increasing its energy expenditures simply to maintain current crop yields. The Green Revolution is becoming bankrupt.

Fossil Fuel Costs

Solar energy is a renewable resource limited only by the inflow rate from the sun to the earth. Fossil fuels, on the other hand, are a stock-type resource that can be exploited at a nearly limitless rate. However, on a human timescale, fossil fuels are nonrenewable. They represent a planetary energy deposit which we can draw from at any rate we wish, but which will eventually be exhausted without renewal. The Green Revolution tapped into this energy deposit and used it to increase agricultural production.

Total fossil fuel use in the United States has increased 20-fold in the last 4 decades. In the US, we consume 20 to 30 times more fossil fuel energy per capita than people in developing nations. Agriculture directly accounts for 17% of all the energy used in this country.[14] As of 1990, we were using approximately 1,000 liters (6.41 barrels) of oil to produce food of one hectare of land.[15]

In 1994, David Pimentel and Mario Giampietro estimated the output/input ratio of agriculture to be around 1.4.[16] For 0.7 Kilogram-Calories (kcal) of fossil energy consumed, U.S. agriculture produced 1 kcal of food. The input figure for this ratio was based on FAO (Food and Agriculture Organization of the UN) statistics, which consider only fertilizers (without including fertilizer feedstock), irrigation, pesticides (without including pesticide feedstock), and machinery and fuel for field operations. Other agricultural energy inputs not considered were energy and machinery for drying crops, transportation for inputs and outputs to and from the farm, electricity, and construction and maintenance of farm buildings and infrastructures. Adding in estimates for these energy costs brought the input/output energy ratio down to 1.[17] *Yet this does not include the energy expense of packaging, delivery to retail outlets, refrigeration or household cooking.*

In a subsequent study completed later that same year (1994), Giampietro and Pimentel managed to derive a more accurate ratio of the net fossil fuel energy ratio of · agriculture.[18] In this study, the authors defined two separate forms of energy input: Endosomatic energy and Exosomatic energy. Endosomatic (MUSCLE) energy is generated through the metabolic transformation of food energy into muscle energy in the human body. Exosomatic (NON-MUSCLE) energy is generated by transforming energy outside of the human body, such as burning gasoline in a tractor. This assessment allowed the authors to look at fossil fuel input alone and in ratio to other inputs...

As an example, a small gasoline engine can convert the 38,000 kcal in one gallon of gasoline into 8.8 kWh

(Kilowatt hours), which equates to about 3 weeks of work for one human being.[19]

In their refined study, Giampietro and Pimentel found that 10 kcal of non-muscle energy are required to produce 1 kcal of food delivered to the consumer in the U.S. food system. This includes packaging and all delivery expenses, *but excludes household cooking).*[20] *The U.S. food system consumes ten times more energy than it produces in food energy.* This disparity is made possible by nonrenewable fossil fuel stocks.

. . . the current U.S. *daily* diet would require nearly *three weeks* of muscle labor per capita to produce.

Quite plainly, as fossil fuel production begins to decline within the next decade, there will be less energy available for the production of food.

Soil, Cropland and Water

Modern intensive agriculture is unsustainable. Technologically-enhanced agriculture has augmented soil erosion, polluted and overdrawn groundwater and surface water, and even (largely due to increased pesticide use) caused serious public health and environmental problems. Soil erosion, overtaxed cropland and water resource over-draft in turn lead to even greater use of fossil fuels and hydrocarbon products. More hydrocarbon-based fertilizers must be applied, along with more pesticides; irrigation water requires more energy to pump; and fossil fuels are used to process polluted water.

It takes 500 years to replace 1 inch of topsoil.[21] In a natural environment, topsoil is built up by decaying plant matter and weathering rock, and it is protected from erosion by growing plants. In soil made susceptible by agriculture, erosion is reducing productivity up to 65% each year.[22] Former prairie lands, which constitute the bread basket of the United States, have lost one half of their topsoil after farming for about 100 years. *This soil is eroding 30 times faster than the natural formation rate.*[23] Food crops are much hungrier than the natural grasses that once covered the Great Plains. As a result, the remaining topsoil is increasingly depleted of nutrients. Soil erosion and mineral depletion removes about $20 billion worth of plant nutrients from U.S. agricultural soils every year.[24] *Much of the soil in the Great Plains is little more than a sponge into which we must pour hydrocarbon-based fertilizers in order to produce crops.*

Every year in the U.S., more than 2 million acres of cropland are lost to erosion, salinization and water logging. On top of this, urbanization, road building, and industry claim another 1 million acres annually from farmland.[25] Approximately three-quarters of the land area in the United States is devoted to agriculture and commercial forestry.[26] The expanding human population is putting increasing pressure on land availability. Incidentally, only a small portion of U.S. land area remains available for the solar energy technologies necessary to support a solar energy-based economy. The land area for harvesting biomass is likewise limited. For this reason, the development of solar energy or biomass must be at the expense of agriculture.

Modern agriculture also places a strain on our water resources. Agriculture consumes fully 85% of all U.S. freshwater resources.[27] Overdraft is occurring from many surface water resources, especially in the west and south. The typical example is the Colorado River, which is diverted to a trickle by the time it reaches the Pacific. Yet surface water only supplies 60% of the water used in irrigation. The remainder, and in some places the majority of water for irrigation, comes from ground water aquifers. Ground water is recharged slowly by the percolation of rainwater through the earth's crust. Less than 0.1% of the stored ground water mined annually is replaced by rainfall.[28] The great Ogallala aquifer that supplies agriculture, industry and home use in much of the southern and central plains states has an annual overdraft up to 160% above its recharge rate. The Ogallala aquifer will become unproductive in a matter of decades.[29]

We can illustrate the demand that modern agriculture places on water resources by looking at a farmland producing corn. A corn crop that produces 118 bushels/acre/year requires more than 500,000 gallons/acre of water during the growing season. The production of 1 pound of maize requires 1,400 pounds (or 175 gallons) of water.[30] Unless something is done to lower these consumption rates, modern agriculture will help to propel the United States into a water crisis.

In the last two decades, the use of hydrocarbon-based pesticides in the U.S. has increased thirty-three-fold, yet each year we lose more crops to pests.[31] This is the result of the abandonment of traditional crop rotation practices. Nearly 50% of U.S. corn land is grown continuously as a monoculture.[32] This results in an increase in corn pests, which in turn requires the use of more pesticides. Pesticide use on corn crops had increased one-thousand-fold even before the introduction of genetically engineered, pesticide resistant corn. However, corn losses have still risen four-fold.[33]

Modern agriculture requires more and more fossil fuel input to pump irrigation water, to replace nutrients, to provide pest protection, to remediate the environment and simply to hold crop production at a constant. Yet this necessary fossil fuel input is going to crash headlong into declining fossil fuel production.

U.S. Consumption

In the United States, each person consumes an average of 2,175 pounds of food per person per year. This provides the U.S. consumer with an average daily energy intake of 3,600 Calories. The world average is 2,700 Calories per day.[34] Fully 19% of the U.S. caloric intake comes from fast food. Fast food accounts for 34% of the total food consumption for the average U.S. citizen. The average citizen dines out for one meal out of four.[35]

One third of the caloric intake of the average American comes from animal sources (including dairy products), totaling 800 pounds per person per year. This diet means that U.S. citizens derive 40% of their calories from fat— nearly half of their diet.[36]

Americans are also grand consumers of water. As of one decade ago, Americans were consuming 1,450 gallons/ day/capita (g/d/c), with the largest amount expended on agriculture. Allowing for projected population increase, consumption by 2050 is projected at 700 g/d/c, which hydrologists consider to be minimal for human needs.[37] This is without taking into consideration declining fossil fuel production.

To provide all of this food requires the application of 0.6 million metric tons of pesticides in North America per year. This is over one fifth of the total annual world pesticide use, estimated at 2.5 million tons.[38] Worldwide, more nitrogen fertilizer is used per year than can be supplied through natural sources. Likewise, water is pumped out of underground aquifers at a much higher rate than it is recharged. And stocks of important minerals, such as phosphorus and potassium, are quickly approaching exhaustion.[39]

Total U.S. energy consumption is more than three times the amount of solar energy harvested as crop and forest products. The United States consumes 40% more energy annually than the total amount of solar energy captured yearly by all U.S. plant biomass. Per capita use of fossil energy in North America is five times the world average.[40]

Our prosperity is built on the principal of exhausting the world's resources as quickly as possible, without any thought to our neighbors, all the other life on this planet, or our children.

Population & Sustainability

Considering a growth rate of 1.1% per year, the U.S. population is projected to double by 2050. As the population expands, an estimated one acre of land will be lost for every person added to the U.S. population. Currently, there are 1.8 acres of farmland available to grow food for each U.S. citizen. By 2050, this will decrease to 0.6 acres. 1.2 acres per person is required in order to maintain current dietary standards.[41]

Presently, only two nations on the planet are major exporters of grain: the United States and Canada.[42] By 2025, it is expected that the U.S. will cease to be a food exporter due to domestic demand. The impact on the U.S. economy could be devastating, as food exports earn $40 billion for the U.S. annually. More importantly, millions of people around the world could starve to death without U.S. food exports.[43]

Domestically, 34.6 million people are living in poverty as of 2002 census data.[44] And this number is continuing to grow at an alarming rate. Too many of these people do not have a sufficient diet. As the situation worsens, this number will increase and the United States will witness growing numbers of starvation fatalities.

There are some things that we can do to at least alleviate this tragedy. It is suggested that streamlining agriculture to get rid of losses, waste and mismanagement might cut the energy inputs for food production by up to one-half.[45] In place of fossil fuel-based fertilizers, we could utilize livestock manures that are now wasted. It is estimated that livestock manures contain 5 times the amount of fertilizer currently used each year.[46] Perhaps most effective would be to eliminate meat from our diet altogether.[47]

Mario Giampietro and David Pimentel postulate that a sustainable food system is possible only if four conditions are met:

1. Environmentally sound agricultural technologies must be implemented.
2. Renewable energy technologies must be put into place.
3. Major increases in energy efficiency must reduce [non-muscle] energy consumption per capita.
4. Population size and consumption must be compatible with maintaining the stability of environmental processes.[48]

Providing that the first three conditions are met, with a reduction to less than half of the [non-muscle] energy consumption per capita, the authors place the maximum population for a sustainable economy at 200 million.[49] Several other studies have produced figures within this ballpark (Energy and Population, Werbos, Paul J. http://www.dieoff.com/page63.htm; Impact of Population Growth on Food Supplies and Environment, Pimentel, David, et al. http://www.dieoff.com/page57.htm).

Given that the current U.S. population is in excess of 292 million,[50] that would mean a reduction of 92 million. *To achieve a sustainable economy and avert disaster, the United States must reduce its population by at least one-third.* The black plague during the 14th Century claimed approximately one-third of the European population (and more than half of the Asian and Indian populations), plunging the continent into a darkness from which it took them nearly two centuries to emerge.[51]

None of this research considers the impact of declining fossil fuel production. The authors of all of these studies believe that the mentioned agricultural crisis will only begin to impact us after 2020, and will not become critical until 2050. The current peaking of global oil production (and subsequent decline of production), along with the peak of North American natural gas production will very likely precipitate this agricultural crisis much sooner than expected. Quite possibly, a U.S. population reduction of one-third will not be effective for sustainability; the necessary reduction might be in excess of one-half. And, for sustainability, global population will have to be reduced from the current 6.32 billion people[52] to 2 billion—a reduction of 68% or over two-thirds. The end of this decade could see spiraling food prices without relief. And the coming decade could see massive starvation on a global level such as never experienced before by the human race.

Three Choices

Considering the utter necessity of population reduction, there are three obvious choices awaiting us.

We can-as a society-become aware of our dilemma and consciously make the choice not to add more people to our population. This would be the most welcome of our three options, to choose consciously and with free will to responsibly lower our population. However, this flies in the face of our biological imperative to procreate. It is further complicated by the ability of modern medicine to extend our longevity, and by the refusal of the Religious Right to consider issues of population management. And then, there is a strong business lobby to maintain a high immigration rate in order to hold down the cost of labor. Though this is probably our best choice, it is the option least likely to be chosen.

Failing to responsibly lower our population, we can force population cuts through government regulations. Is there any need to mention how distasteful this option would be? How many of us would choose to live in a world of forced sterilization and population quotas enforced under penalty of law? How easily might this lead to a culling of the population utilizing principles of eugenics?

This leaves the third choice, which itself presents an

unspeakable picture of suffering and death. Should we fail to acknowledge this coming crisis and determine to deal with it, we will be faced with a die-off from which civilization may very possibly never revive. We will very likely lose more than the numbers necessary for sustainability. Under a die-off scenario, conditions will deteriorate so badly that the surviving human population would be a negligible fraction of the present population. And those survivors would suffer from the trauma of living through the death of their civilization, their neighbors, their friends and their families. Those survivors will have seen their world crushed into nothing.

The questions we must ask ourselves now are, how can we allow this to happen, and what can we do to prevent it? Does our present lifestyle mean so much to us that we would subject ourselves and our children to this fast approaching tragedy simply for a few more years of conspicuous consumption?

Author's Note

This is possibly the most important article I have written to date. It is certainly the most frightening, and the conclusion is the bleakest I have ever penned. This article is likely to greatly disturb the reader; it has certainly disturbed me. However, it is important for our future that this paper should be read, acknowledged and discussed.

I am by nature positive and optimistic. In spite of this article, I continue to believe that we can find a positive solution to the multiple crises bearing down upon us. This article is simply a factual report of data and the obvious conclusions that follow from it.—DAP

Post Script

Consider this from a Reuters news story on October 3, 2008, the day the $700 billion (plus $150 billion in add-ons) bailout bill for Wall Street was signed by President Bush.

"Farmers also face a tighter squeeze ahead as fertilizer costs have tripled in the past two years while seed and fuel costs have doubled . . .

". . . [John Conway] added that his farmer cooperative borrowed more money this year for a single input, anhydrous ammonia fertilizer, than it did for all operating costs combined two years ago. The crucial fertilizer has risen from $350 a ton in 2006 to near $1,200 . . .

"If credit is unavailable, farmers may plant fewer acres or cut back on fertilizer and hope their fields have enough residual nutrients to carry another crop, analysts said."[53]

RECOMMENDED READING

http://www.permatopia.com/
http://www.holmgren.com.au/

CHAPTER NINE

Evaluating Alternative Energies

This chapter deals with the harder questions about all of the alternative energies that are so hotly touted in the media by politicians and major corporations—the ones that so far have not been asked or answered for the public. The time has come when they must be. It has been said several times that governments cannot make policy or govern based solely upon hope any more than he or she can make decisions based upon a corporate press release. At some point the nuts and bolts of alternative energies have to be looked at. One has to call in the engineers, the financial experts, and the scientists and ask, "OK, how do we put this together and make it work?"

Can we make it work?

Assume you are the president and it is now your job to pick and choose between all the alternatives that seem so promising.

I have been studying this problem for seven years, reading hundreds of thousands of words, visiting thirteen nations and asking many questions. The truth is there is no alternative energy, or combination of alternative energies, that will permit current consumption and lifestyles to continue—let alone provide for the compound growth we are wedded to in the current economic paradigm. We have already examined the monumental infrastructure issues that must be addressed. We have seen that instead of addressing these critical issues thirty, twenty, or even ten years ago, with a rational and organized plan, the United States and much of the world, following the U.S. example, has stalled, delayed and denied. Our culture of consumption, wealth and good times pervades the world's mindset through movies, TV and advertising. It's easy for us to portray and market this mindset.

We are five percent of the world's people using 25 percent of the world's energy production. We make it look easy, but we have neglected to acknowledge that it's not possible for everybody to live this way.

An American president can function effectively only in partnership with the people, Congress, the Supreme Court, and business. He or she cannot materialize hamburgers while locked inside a vault of public expectations or change the laws of physics and thermodynamics just because people demand it.

A "let them eat cake" approach will not solve anything. A bloody revolution started just after those infuriating words were allegedly spoken by Marie Antoinette.

After maybe fifty lectures with lengthy question-and-answer sessions I, and almost every other sustainability advocate who has worked so hard for years to start a rational discussion, can almost sleepwalk through the questions from the audience. They all begin with, What about . . . ? This is the biggest hindrance to finding workable solutions and the first stage of grief. It is denial. It is present not only in the general population but in the mainstream media, which constantly feed an endless pep rally promoting what alternative energy sources are supposed to be able to do without subjecting these claims to the most rudimentary critical analysis . . . or what they learned in high school science classes. It is present in Congress, where leadership from both sides refuses to think critically and confuses almost everything.

With a few bright exceptions like Congressman Roscoe Bartlett (R-Maryland), who has diligently tried to educate Congress on the true magnitude of this crisis (e.g. the issues of Peak Oil, energy shortages, compound growth and the breakdown of society), those elected to protect our interests stay clear of these non-starters for politicians. Who gets re-elected by offering bad news? Challengers can be optimistic without ever having to be right because memories have been short and no one has yet had the will to hold Congress or a president truly accountable for their many failures of leadership. The problem could always be passed on.

An American president who tries to come to grips with the energy crisis will run headlong into an ingrained American attitude which says that every problem can be solved by throwing money at it. But that cycle will end before the 2012 U.S. presidential election when the accounts come due and can no longer be passed on—when printing more money will not solve the crisis.

Money has no value without energy to back it up. Problems *can* be solved by throwing energy at them. But what kinds of energies?

Feedback loops are now so short that those who have been elected, or who presently hold office, will be remembered for ineptitude, vacuous promises and lack of vision. Media outlets that have presented lies and bad information as fair and balanced against the truth will be shunned and boycotted in short order because of their failures. The people will remember what they were told.

Following the famous utterance attributed to Marie Antoinette came the French Revolution and Reign of Terror, during which centuries of pent-up rage vented violently, producing chaos, uncertainty and paralysis that led ultimately to a dictatorship in France and an empire that engulfed the world in war.

If one truly loves this country, this and the next American president must look beyond to the leadership of Abraham Lincoln after the Civil War: "With malice towards none; with charity for all." To mix metaphors, the "Ancien Regime" will pass and must. Those who represent the old paradigm must be removed from positions of authority and replaced. But in what manner? Revenge and war should have no place in a president's future planning, no matter how much a very predictable human nature may seek it.

War is the greatest waste of energy and resources there is. It purposefully destroys things which took enormous amounts of energy to make—so that they can be made again . . . and destroyed again . . . and made again . . .

So let us collectively sit down behind the desk in the Oval Office and look at things the way a president might at the moment when the energy crisis cannot be postponed or evaded any longer. That moment is, of course, right now.

Presidential Questions About Alternative Energy

Before we instantly accept alternative energy lifeboats that will allegedly let us keep our current lifestyles, don't you think it wise to see if they float? We would expect our leaders to do that. Before committing limited resources to address the crisis, shouldn't we set some standards to evaluate each one? Should we not ask questions about equity, fairness, and what will best and most efficiently serve the nation and its people as a whole? And, if it isn't possible to please everyone, how do we prioritize our allocations and make the most difficult choices of all?

Shouldn't a president take a step back from pork-barrel politics as usual and the lobbyists with campaign donations promoting specific industries, shareholders and special interests? Can we afford a Green Energy bubble that will work about as well as the Housing or the Dot-com bubbles worked? The markets have failed miserably to prepare us. They have no medium- or long-term vision, and their only priority is to benefit shareholders.

Reality

The first reality that will not yield to a presidential order or opinion poll is the science of physics which, from an energy standpoint, is governed by the Laws of Thermodynamics.

1st Law—Energy can be changed from one form to another, but it cannot be created or destroyed. The total amount of energy in the universe remains constant, merely changing from one form to another.

2nd Law—In all energy exchanges, if no energy enters or leaves the system, the potential (usable) energy of the state will always be less than that of the initial state. Energy only converts in one direction, from useable to unusable. Big things break down and get smaller until they reach a point of stasis or balance. This is also known as the law of entropy.

3rd Law—It is impossible to cool a body to absolute zero by any finite process. This is actually more of a postulate than a law. In any case, it has little application to our discussion and is presented here merely for thoroughness. (Note: Absolute Zero is the theoretical point at which all atomic motion stops or freezes.)

Scientist and author C.P. Snow developed a very simple and memorable way to remember the three laws:

- You cannot win (that is, you cannot get something for nothing, because matter and energy are conserved).
- You cannot break even (you cannot return to the same energy state, because there is always an increase in disorder; entropy always increases).
- You cannot get out of the game (because absolute zero is unattainable).

Here are some questions our leaders must ask of anyone who claims that they have found a perfect alternative or combination of alternatives to oil, coal, and natural gas. After answering these questions, you may have a better idea about whether you want to jump (or throw your family and your nation) into something that might sink just as soon as their weight bears down upon it. There is a vast difference between temporary and permanent solutions here. A president cannot mistake prudent, but ultimately stopgap, measures that must be taken to soften the impact of the crisis as *solutions* to the crisis.

After answering all these questions, you will see—from a scientific perspective, rather than an emotional one—that there are no effective replacements (or combination of replacements) for what hydrocarbon energy provides today. This leaves the president with the decision of supporting those choices that have the best chance of allowing the nation to function until a new, greatly slimmed-down energy and economic regime can evolve and, most importantly, be put in place; one that will inevitably require greatly reduced production, consumption and waste across the board.

In fact, this "Power Down" option is already well underway. While selling alternative energies as a means to increase energy supply, politicians, investment banks, and Wall Street are pursuing an entirely different strategy as witnessed by the current depression and "demand destruction" taking place. These decision-makers, though promoting rosy optimism, have already answered these questions in private. This reality was excellently described by Professor Richard Heinberg in his 2004 book of the same name, *Power Down*.

The bottom line is that if supply cannot be increased, then demand must be decreased. And the only way to decrease demand is through the curtailment of economic activity. What is being pursued through alternative energies is not an increase in energy supply; it is an *inadequate* form of energy substitution that cannot permit our lives to continue as they have. While in free fall, the powers that be are attempting to construct a parachute out of disparate and mismatching parts that serve the monetary system first and people last.

These questions will help us pick and choose between the options.

1. How much energy is returned for the energy invested (EROEI)? (Also called Net Energy)

Historically, in industrial civilization the energy returned for energy invested has been enormously high. Oil and natural gas are two of the most concentrated forms of energy ever discovered. There is no fixed EROEI for oil or natural gas or any other energy source. When oil was first used commercially in Pennsylvania it was literally seeping from the ground. No need to invest much energy there. Contrast that with wells drilled tens of thousands of feet (i.e. miles) down into the earth using oil-powered drills and deepwater drilling that takes many times the energy investment of land-based drilling. All the energy costs of building the rig, towing it out to sea, and then drilling two to three times deeper than on land must be considered. If oil is found, the energy costs of pumping and transporting it must be taken into account. And then what happens if your well is a dry hole, as is more and more often the case?

In the case of natural gas, drilling into a pocket that contains 500 million cubic feet returns a lot of energy for relatively little investment. But, since we have plucked all the low hanging, big fruit, we are now drilling into pockets of only 50,000 or 100,000 cubic feet, or less. When that pocket runs dry a new well must be drilled and the energy return recalculated on that basis. New wells are typically running dry in a year or less.

The Texas Railroad Commission (TRRC) has historically been the repository of statistics and information on the number of wells drilled for both oil and natural gas. Texas' experience mirrors that of North America and the planet as a whole. America's oil and gas industry began to organize, regulate, and track itself in Texas during the early years of the last century when the oil boom began. Since everything was shipped by rail car, that was the perfect place to start keeping track. As much as Texans like to think of themselves and their state as unique, Texas does not operate under different laws of physics than the rest of the universe—though some might argue the point.

A check of TRRC's statistics shows oil production (as we already knew) in a steady decline, but natural gas production has been increasing year after year from 2002 through 2008.[1] Sounds great, especially for people like billionaire oilman turned Peak-Oil-solver T. Boone Pickens, who advocates switching much of our vehicle fleet to natural gas. Burning natural gas in cars emits fewer greenhouse gases than oil, but the notion that it is clean is about as valid as "clean coal" being clean. That's an improvement for environmentalists and all of us. There's a big question, though, about who's going to pay to convert all our gasoline-burning engines to natural gas. Can you afford a couple thousand dollars to convert your car right now? You might not want to when you see what's next.

There is a deal-killing devil lurking deeper in the details.

Consider that in 2002 a total of 10,966 new gas wells were drilled in Texas. In 2004, 13,665 new wells were drilled.[2] That's an increased drilling rate of 2700 new wells per year in just two years. Older data archived on the TRRC web site reveals the stark truth. In 1960 the state of Texas completed 2,011 new natural gas wells. In 2007 it completed 8,643.[3]

That means that smaller pockets are being found and exhausted more quickly. In order to keep pace with growing demand, more money has to be spent to find increasingly smaller pockets of gas. Then there is the surprise factor of natural gas depletion. Unlike oil, gas flows around geological obstacles. So when a pocket is dry, there is often very little warning before production drops off, just the way a balloon will rapidly deflate until all the air is gone and then just stops.

EROEI is at the heart of what sustainability means. This is also what will allow presidents and planners to equate apples and oranges when it comes to the benefits of each proposed alternative. Have all energy costs been taken into account? This is where too many popular alternative energy sources fall flat in the wake of the simplest examination. Public discussions of these alternatives generally ignore the question altogether, preferring to leave a consumer believing in free lunches while spending more on non-essential items.

Commercial hydrogen offers one clear example of how it takes more energy to produce the fuel than can be obtained from burning it. The current feedstock from which hydrogen is produced is natural gas. Natural gas is treated with steam. Steam is produced from water that is boiled by using more natural gas, oil, coal, or nuclear energy in the form of direct fuel or indirectly generated electricity. Common sense dictates that this cannot be a solution, because it still relies almost exclusively on fossil fuels. We have already seen how a 1999 University of California study revealed

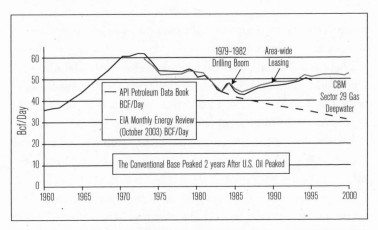

U.S. Gas Production Actually Peaked in 1973! SIMMONS & COMPANY INTERNATIONAL. TAKEN FROM *THE NATURAL GAS RIDDLE: WHY ARE PRICES SO HIGH? IS A SERIOUS CRISIS UNDERWAY?* SIMMONS, MATTHEW. HTTP://WWW.SIMMONSCO-INTL.COM/FILES/IAEE%20 MINI%20CONF.PDF

that more than 1,113 gallons of gaseous hydrogen is necessary to produce the same energy as a gallon of gasoline. Many people ask about building nuclear reactors to generate steam. We will deal with nuclear energy in a separate section.

Converting water to hydrogen is done through electrolysis, or the application of electrical energy. Scientist David Pimentel has established that it takes 1.3 billion kWh (Kilowatt hours) of electricity to produce the equivalent of 1 billion kWh of hydrogen.[4] In other words, by this method, hydrogen is a net-energy loser . . . a waste of energy.

Even a small positive EROEI, if obtainable, is not a solution because fossil fuels on the whole have always returned sometimes hundreds of times the energy invested, not just a small fractional increase. That's why we used and depended upon them for a century—and stopped thinking about alternatives until it was too late. Oil can only be burned once.

Ethanol is another case in point. Some excellent research has shown a negative EROEI for ethanol. In other words, it takes more energy to make ethanol than you get from burning it. Newer research shows a slightly positive EROEI in some cases. All that was lost was a lot of food, a lot of nutrients in ever-depleting soils, a lot of time, a lot of oil and gas—and *lots and lots* of fresh water.

Any alternative energy source claiming to be a solution must have documented "open book" EROEI policies. If it doesn't, then it has something to hide.

Solar technology is improving rapidly and will be a vital part of our future. As of about 2006, however, state-of-the-art solar production was extremely energy intensive. I have seen studies showing that it took five years for one solar panel to generate as much energy as that required to make it. That is consistent with news stories reporting that electrical ratepayers were waiting that long, or longer, before their original cash investment paid for itself in savings.

As this book was being finished, there were new reports of breakthroughs that dramatically simplified solar energy production by making a spray-on "solar film." That is fabulous news. The only problem is that we should have had that process thirty years ago. It seems clear that many cars in the human future will be electric, charged at home from solar panels and household batteries that are practical and involve no long-distance transmission. But there will never be a billion—or even 500 million—electric vehicles with seven gallons of oil in each of four tires and oil in each piece of plastic to save weight.

The biggest villains on EROEI are tar sands, so-called shale oil, coal-to-oil, and ethanol.

2. Is the energy regime a substitute or just scavenging?
Solar, wind and tidal energies are true replacements for oil and natural gas *where they are feasible*. Unfortunately they have severe limitations as this chapter demonstrates. They only generate electricity which will not power a 2008 GM hybrid—for long, that is.

The fact that diesel engines can run on used vegetable oil never takes into account the amount of energy necessary to generate the vegetable oil in the first place (farming, vegetable transport, extraction, etc.). However, in contrast to other biofuels, the waste vegetable oils are essentially useless for cooking by the time Burger King throws them out and can be converted in a simple, layman-friendly process to biodiesel. No one has ever suggested, however, that Americans should or could eat enough French fries, battered fish, fried chicken and shrimp to power America's diesel fleet. It might, however, be a good way to reduce population.

Recycling vegetable oil or converting other food and plant waste that cannot be returned to the soil to make diesel are just means of scavenging the leftovers of our fossil-fuel extravagance. These methods will be essential to the survival of human civilization but they cannot be considered substitutes.

Devices that recycle plastic into oil don't mention the fact that plastic *is* oil, and that a great deal of energy was used to make it into plastic in the first place. The logic of converting oil into plastic and then back to oil is about as sensible as converting oil and natural gas into food and then back into gasoline. So any energy policy promoting biodiesel or the recycling of anything should be labeled clearly in public and official consciousness as a band-aid on an infected wound rather than a cure.

Similarly, the new technology of thermal depolymerization is not a legitimate alternative energy source. This process transforms carbon-based wastes back into hydrocarbon fuel. This technology is useful and may help us on the downside of the Peak Oil curve, but it will never replace fossil fuels. Why? Because the wastes were produced by the use of fossil fuels in the first place, and the Second Law of Thermodynamics tells us that energy is always lost in an energy transfer or exchange. I remember an article touting thermal depolymerization, saying that one could throw a 170-pound man in one end of the machine (powered by electricity) and extract x liters of oil on the other end. Great! Soylent Oil.

As fossil fuels dwindle, so will the source materials—even humans.

3. Have the claims been verified by an independent third party?
In real life, it is called "the proof is in the pudding." In scientific circles, it is called peer review, and it usually involves having research published in a peer-reviewed journal. It is an often-frustrating process, but peer-reviewed articles ensure the validity of science.

When assessing the validity of an alternative energy source, look for articles published in peer-reviewed scientific journals or critiques authored by scientists or engineers trained in the field. Ultimately, this is the only way to validate claims. An inventor may insist that he or she has been shunned by the scientific community or that there is a conspiracy within the scientific community against his or her ideas. That is just too bad. Don't succumb out of wishful thinking.

The ultimate proof is in a working demonstration outside of the control of the person selling the idea so that the results can be verified by a person or body with no conflict of interest. Hydrogen cars can and have been made, but they are the worst form of fraudulent advertising I have ever seen. Each car is an example of waste. Who cares if water comes out of the tailpipe instead of greenhouse gases? What we don't see is the fact that not one of these vehicles is commercially viable, that there's no place to fill them up, or that fuel cells wear out quickly and need to be replaced (more hydrocarbon energy used). Automotive advertising has implied that hydrogen cars will be here any second. The fine print acknowledges that they won't be a viable option for maybe thirty years after certain "technical problems" are solved (i.e. the laws of thermodynamics are somehow overturned).

Another great writer on the subject of Peak Oil (and gas) is James Howard Kuntsler. In May of 2006, while I was attending (and speaking at) a Peak Oil conference at New York City's Cooper Union, I saw him show a slide of a 747 pulling up to a filling station. The pilot rolled back his window and yelled out, "Fill it up with technology!"

Doh!

4. Is the energy available 24/7 as needed?

Windmills only generate electricity when the wind blows. Solar panels only generate electricity when the sun is shining. There is no way to store electricity in sufficient quantities to meet U.S. needs. No one has ever suggested enough giant batteries (using technologies that do not exist) be made from enormous amounts of energy to save electricity for a calm day or a dark night. How often does one have to change batteries in a flashlight and throw them away? Lithium batteries and rechargeable batteries are horribly energy intensive to make, toxic, and, as we all know, very expensive just for a camera, let alone a house or a neighborhood. What about Philadelphia? What about the waste?

5. Is the energy transportable over distance?

One of the great frequently unmentioned beauties of oil and gas is that they can be easily transported for use where needed via pipelines or tanker vessels. This is a major problem for electricity for two reasons. First, the laws of physics state that electricity will be drawn off first and used closest to where it is generated. That is why a wind farm off the coast of, for example, Martha's Vineyard in Massachusetts, will be of great value to the upper-class families who live there. Most of the electricity generated will be long used before it gets to the working-class neighborhoods of Boston. The same holds true for wind corridors. Build thousands of windmills in Texas, Oklahoma, Colorado, and Nebraska and the primary beneficiaries of that will be Texans, Oklahomans, Coloradans, and Nebraskans.

Yet there are proposals to use ever-shrinking federal tax dollars to do just that. That would amount to having some folks pay for other folks' energy. It would also amount to people in one state paying for electricity in other states far away.

6. Is the energy source applicable for the region?

Geothermal energy derived from cracks in the earth's surface may be great for Iceland, where it is readily accessible and abundant, but that has nothing to do with New York City. The abundant solar energy in California's desert does little good for the Pacific Northwest. And no one has yet answered where hundreds of thousands of miles of new transmission lines and transformers are going to come from. No one has even asked if there's enough copper left to be mined, destroying more arable land, to make the transmission wires. Remember, it takes thirty years to change an energy infrastructure and billions (if not trillions) of dollars.

This one question will lead us into what is the brightest spot in what by now must seem a horribly depressing situation. Whatever solutions emerge to the energy crisis will be local, rather than national. And it will be the greatest task of an American president to use the resources of our federal government to facilitate local preparation and response to a crisis that will, on almost all levels, be met by counties, cities, towns and even neighborhoods and households.

This will be the subject of a coming chapter.

7. Does the inventor claim zero pollution?

There is no method of generating energy from a source that does not produce some form of waste (pollution). Even wind and solar create waste as a result of the construction of wind turbines and solar cells (albeit comparatively little waste, generated largely in their initial manufacture). Hydrogen fuel cells create waste when the hydrogen is generated, though it is commonly claimed that they produce nothing but water. They depend on fossil fuels to generate the free hydrogen, so they create all the pollutants of burning hydrocarbons; they simply move them away from the vehicles to a centralized generating plant. All of it winds up, however, in the same ecosystem we call Earth.

Likewise, horses produce waste; just ask anyone who has ever mucked a stable. As a matter of fact, at the beginning of the twentieth century the streets of New York City were literally buried under horse manure. By the way, it takes about five acres of arable land to support one horse. We could do that in 1900 when the population was around a hundred million people. Do it now and there won't be any land left to grow food on. Oh yes, I forgot. We need all that land for mono-cropped ethanol anyway.

Who needs to eat?

Electric vehicles pose the same problem. Electricity is not an energy source. It must be generated. Currently the U.S. gets about 50% of its electricity from coal and roughly 20% each from natural gas and nuclear. Hydroelectric, oil, renewables (wind and solar), and other gases (e.g. propane) make up the remaining 10%. According to the Department of Energy (DoE), 90% of all new electrical generating plants to be built in the next two decades will be gas-fired. All that clean electricity that would now go into powering an electric car is simply overlooking the greenhouse gases emitted by the coal-fired plant that produced it.

8. How destructive of the environment is this energy source?
Rarely do the media or Congress or the Department of Energy go directly at this topic, especially when it comes to subjects like coal-bed methane, oil shale, or tar sands.

Going back to ethanol again, where large amounts of energy are used to produce steam, here's a recent quote from the web site *AlterNet*:

> That's not all. "[Ethanol] plants themselves—not even the part producing the energy—produce a lot of air pollution," says Mike Ewall, director of the Energy Justice Network. "The EPA (U.S. Environmental Protection Agency) has cracked down in recent years on a lot of Midwestern ethanol plants for excessive levels of carbon monoxide, methanol, toluene, and volatile organic compounds, some of which are known to cause cancer."[5]

Shale Gas—A Magic Bullet?
In the summer of 2008, financial stories started celebrating a new

rise in natural gas production and a resulting drop in natural gas prices. The source of this apparent reprieve was so-called "shale gas" where natural gas is liberated from shale formations by cracking ("fracking") rock and injecting water and other chemicals to open and pressurize flows to wellheads.

While I was able to locate a number of stories touting the fact that shale gas production had increased U.S. gas reserves dramatically and boasting of perhaps hundreds of trillions of cubic feet of new supply—literally a bonanza of new energy—the story was too new to draw hard conclusions. No one has yet issued figures as to how much of this may be recoverable. Almost nothing has been formally said about what the environmental or monetary costs may be.

It is clear that U.S. domestic gas production rose dramatically in the summer of 2008. The *New York Times* reported that, in contrast to years of consistent declines, "domestic gas production was up 8.8% in the first five months of this year compared with the period a year earlier, a rate of increase last seen in 1959. . . . " [6] It is also clear that natural gas prices decoupled from oil prices, falling much faster and farther than oil prices after the oil price spikes of June and July 2008.

The largest shale gas play in the country is the Barnett shale field located near Dallas—Fort Worth. Other large plays are located in Arkansas and in Appalachia, with additional fields in the Catskills and other parts of the country. A *New York Times* story on August 25, 2008, quoted the chairman of Chesapeake Energy as saying, "It's almost divine intervention." The same story also disclosed that House Speaker Nancy Pelosi had invested in a company that produces natural gas for automobiles. [7] (Natural gas is natural gas. There's no special kind needed for cars.)

Surprisingly, there was not a lot of serious detail on how shale gas is produced or what environmental risks might be involved. Nobody seemed inclined to disclose what kinds of chemicals were involved exactly. I remember the old adage, "If it sounds too good to be true, it probably is." The shale gas story was just too new as this book was being finished to draw firm conclusions, but some serious warning lights are already flashing.

I found the following from a two-part series posted at www.triplepundit.com in August of 2008. None of the financial or investment stories mentioned any of this.

Part One of the series was titled "Shale Gas: Energy Boon or Environmental Bane."

> . . . *But at What Cost to Water Resources?*
> That's not the whole story, however, according to local residents, reporters and environmentalists. Each shale gas well on average requires some 4.5 million gallons of water, and there can be many wells at a single site.
>
> Hundreds of trucks transport water into drilling sites. They then truck the recovered water, with grit and chemicals mixed in, back out for disposal. How is it disposed of? Well, it's difficult to find an answer and Texas reportedly has no regulations covering it.
>
> Documented reports of contaminated well water are growing, as are indications that the water table in the Fort Worth area is falling. Both are mortal threats to farmers and ranchers, as well as the region's general population, and they're not the only mortal risks associated with shale gas drilling and production, topics which will be explored further next week in Part Two.[8]

Part Two of the series was appropriately called "Boom & Bust, Boon or Bane: Shale Gas Fever Spreads"

> . . . As shale gas fever spreads and drilling and production increases, the environmental costs are becoming increasingly apparent, however. Reports of contaminated water supplies, sinking water tables, explosions and drilling accidents are on the rise, even as shale gas drilling spreads into densely populated urban areas, prompting calls for greater

oversight, regulation and rules to protect neighborhoods and the environment. . . .

Threatened water supplies
The possibility of contaminating New York City drinking water has prompted city officials to demand that the State Department of Environmental Protection ban natural gas drilling and establish a one-mile buffer zone around the Ashokan Reservoir and each of the six major Catskill Mountains' reservoirs spread across five counties, as well as the connecting infrastructure, that supply the city's drinking water, according to an August 6 ProPublica news report. . . .

Reassurances but little or no actual oversight
The New York City officials have good cause for concern, and a growing amount of evidence to support their concerns and calls for more stringent, more comprehensive environmental protection and transparent oversight.

Shale gas drilling and production is highly water-intensive. Liberating natural gas from shale and getting it to flow from depth to the surface requires large, reliable water supplies, which are mixed with drilling mud, grit and chemicals to fracture, or "frack," the tightly compressed shale to facilitate gas flow.

. . . After it's used the recovered, tainted water is trucked out and disposed of, how and where no one outside the oil and gas companies and their contractors apparently knows, and they aren't saying.

Part Two concludes by revealing that no environmental impact studies are required anywhere to produce shale gas.[9]

Quite frankly, shale gas scares me to death and suggests a desperation level I wasn't sure we had reached yet.

In January of 2009 I found a source who had worked in a shale gas field. He told me—on condition of anonymity—that he knew, because he had seen it, that some of the chemicals added to fracking compounds included Boric, Hydrochloric, and Citric acids and bromides.

I hold the reader responsible for looking up these chemicals and determining their potential for damage to water tables and human populations.

9. What are the infrastructure requirements?

Does the energy source require a corporation to produce it? How will it be transported and used? Will it require new engines, pipelines, and filling stations? What will they cost? Who will pay for them and with what? How long will it take to build them?

If a technology is complicated, requiring specially trained technicians, sophisticated machinery, and elaborate processing, then major corporations and/or governments will likely control it. This will leave you with very little say in the matter. You will simply remain a consumer paying your bill or a stockholder collecting your premiums.

Nuclear fast-breeder reactors have excellent net-energy profiles, even better than fossil fuels. But, if they are ever perfected, you can bet that you won't be able to build one in your garage. They will be owned and managed by corporations. The waste is dangerous, and there isn't enough uranium to supply the world's energy needs anyway—not with an exploding population. There is such a thing as Peak Uranium, and we are at that point too.

There are a few technologies that do offer useful net-energy profiles (while not approaching fossil fuels), and are available for home use. Windmills, passive solar (solar heating), and paddle wheels are examples of such technology. Even hydrogen, which I have so thoroughly discounted for large-scale answers, shows some promise for limited applications in individual households and businesses. Methane processing of farm wastes has received some attention (particularly in traditional Asian cultures), but it generally involves some advanced machinery and is potentially dangerous because methane is so highly combustible. None of these technologies are not scalable to a modern industrial civilization; they will work only locally, unless one wishes to argue that pig manure should be shipped by gasoline or diesel from Arkansas to New York to generate electricity.

RECOMMENDED READING

The Party's Over by Richard Heinberg (New Society, 2005).
Power Down by Richard Heinberg (New Society, 2006).

CHAPTER TEN

The Alternatives

Solar

One of the two best—if limited—alternatives. The federal government knows this because it has been moving with relatively little fanfare, but on a massive scale, to convert major military and other key installations to solar power. We documented this extensively in my former newsletter, *From The Wilderness*. But we have already discussed some of solar power's inherent limitations. At this stage of transition it is, in most cases, more of a means to stretch out fossil fuel energy rather than to replace it. It is also very expensive, and the production of solar panels and other equipment is far below the levels needed to begin across-the-board conversion away from other sources. There are long waiting periods for solar panels in most places already.

Solar is an easy choice for a president. Promote it and use the federal government to support it in any way possible, especially with feed-in-tariffs, which will be discussed in the chapter on money. Solar will reduce fossil-fuel use for energy consumption, but will do very little or nothing in the short- and medium-range terms to assist transportation issues or food production.

Wind

Another clear, but limited, choice. It's not quite free because someone has to pay for, build, install and maintain the windmills, transformers, and power lines. It cannot be applied uniformly throughout the country. Infrastructure and capital investment requirements—in terms of time, raw materials, and cost—will not produce the 270 million electric-powered vehicles needed in the U.S. (or the estimated 800 million for the planet) that will plug into windmills in time to soften the landing. (Remember the seven gallons of oil in every tire.)

Tidal Energy

This is something worth looking at. Tides ebb and flow, and this enormous energy generated by the gravitational pull of the moon can be captured to some extent by building large devices that generate electricity in much the same way that dams do. The issues again are cost, development time, and the obvious fact that tidal energy will work only near a coastline. Sorry, Denver. Sorry, St. Louis. Sorry, Chicago. And, there are other problems with this new and untried technology.

Tidal is more expensive than coal or oil. Not all the issues have been worked out with regards to conversion of tidal energy to electricity. Appropriate waves and tides are highly location dependent. Waves are a diffuse energy source, irregular in direction, durability and size and therefore unpredictable. Extreme weather, for example, could produce massive waves that would overwhelm capacity.[1] Talk about popped circuit breakers! To date, no one has found a way to concentrate the widely dispersed energy of tides. Facilities large enough to produce sizeable amounts of electricity might interfere with natural ebb and flow which is essential to sea life. And finally, salt water is corrosive to metal.[2]

Coal And "Clean Coal"

Coal is not strictly an alternative energy. However, it is being increasingly promoted as a substitute for petroleum and gas, so we will discuss it here.

Now our choices get much more difficult. Yes, it is true that coal is America's most abundant energy resource. We have been falling back on coal, even with climate change hanging over our heads, for some time. As of mid-2008 the Department of Energy reported that 52 new coal-fired plants had been permitted, were near construction, or under construction with another 58 in the early stages of development.[3]

As president, you might sit down at your desk with high expectations here. Then reality would set in. The coal industry has
spent millions *promoting* clean coal and bragged about plans to
strip out the principal greenhouse gas, i.e. carbon dioxide, calling
the end result "clean coal." Clean coal is one of the biggest lies
in human history. Barack Obama has committed his administration to clean coal, and this absolutely suicidal policy must be
abandoned.

First you would learn that there is not a single clean-coal generating plant operating anywhere in the country (or the world).
Coal combustion emits many poisonous substances like sulfur
dioxide, nitrogen oxide, and lots of particulate matter in addition to CO_2. Commercially viable carbon sequestration, or CCS
(Carbon Capture and Storage), you would learn next, is still a
theoretical and hugely expensive proposition. It is being done in
real life, but not on a cost-effective basis and only in a few places.
It is not being done at any generating plant anywhere. Even while
it could theoretically reduce CO_2 emissions by 80–90% (from
combustion, but not from mining) at a cost as high as $8 per
ton and huge additional energy requirements to heat and process
the CO_2 out, you would still face the very expensive proposition
of compressing and transporting the CO_2 over long distance to
either underwater caverns or to geologic formations and empty
natural gas wells.[4]

Coal's solid waste—Local newspapers pay close attention to
issues like this because it is localities that have to deal with the
problem. In 2008 a New Mexico newspaper reported this:

> Each year, power plants in the U.S. collectively kick out
> enough of this stuff to fill a train of coal cars stretching from
> Manhattan to Los Angeles and back three and a half times.
> It's stored in lagoons next to power plants, buried in old coal
> mines and sometimes just piled up in the open. It is the larg-
> est waste stream of most power plants, and a recently released
> study by the Environmental Protection Agency found that

people exposed to it have a much higher than average risk of getting cancer. Yet the federal government refuses to classify the waste as hazardous, and has dragged its feet on creating any nationally enforceable standards. And with new attention focused on coal power's impacts on the air, this great big problem may get worse, and continue to be ignored.[5]

All of this drastically impacts EROEI.

Your mood would sink a little more when you were briefed that America's first clean-coal plant was being built in Illinois and scheduled to go online by 2013, but that Bush Energy Secretary Samuel Bodman later reversed the approval decision and withdrew funding. Here's what a newspaper from India (no less) reported:

> COAL-FIRED power plants are taking hits from all sides. The unkindest cut to future coal-fired power generation came recently when Samuel Bodman, Secretary, the US Department of Energy (DOE), declared that the Bush administration had decided to withdraw funding to FutureGen, the US government's effort to develop a "clean coal" power plant.
>
> The plant would have turned coal into hydrogen-rich synthetic gas, generating electricity while pumping carbon dioxide underground for permanent storage (*The Wall Street Journal*, *WSJ*, February 2, 2008). The project had international participation.
>
> The DOE found that the cost of the project soared to $1.8 billion, nearly double the original estimates.[6]

Dan Becker has worn two hats. He is now director of the Sierra Club's Global Warming and Energy Program, which would make him appear a Liberal on the issue. But he is also a former head of the Tennessee Valley Authority (TVA), one of the largest electrical generation operations in the world. Becker said:

There is no such thing as clean coal and there never will be. It's an oxymoron. I say this based on my experience as the former head of the TVA, which bought and burned more than 30 million tons of coal a year. I was deeply involved in the strip mining, underground mining, trucking and most importantly, the burning of huge quantities of coal. No one who has been deeply involved with coal can rightfully say it is clean.[7]

Julian Darley, now an energy analyst based in London, is the founder and past-president of the Post Carbon Institute (http://postcarbon.org/) which has, for many years, been studying Peak Oil and all the implications of energy shortages. As the name implies, his organization believes that the only possible future for mankind will be inevitably be one where carbon-based fuels are a sidebar accessory to other, sustainable regimes. One of his major focuses is to assist communities in relocalizing to move towards self-sufficiency in preparation for the energy-induced collapse that is already underway. The Post Carbon Institute also works with local governments on issues pertaining to energy shortages and transitioning away from carbon-based fuels. As part of his work he and his staff study all of the so-called solutions to the energy crisis, including coal.

In an interview for this book he noted that the process of mining coal by itself produces enormous amounts of toxic waste. Clean coal does not address that end of the equation at all. Then, with regards to carbon sequestration as it has (not) been implemented thus far in coal-fired plants, he observed, "What the plant builders say, once they have the plant built is, 'We've run out of money to implement carbon sequestration. We'll come back and do it later maybe . . . when there's more money."

All those new proposed clean-coal plants are going online as dirty-coal plants.

"When there's more money," they tell us . . .

As president, you look at the state of the economy and the federal budget, the deficits and the national debt, and you understand that there isn't going to be any retrofitting later on. There are bridges to repair, roads to be paved, sewers to be maintained, corporations like Bear Stearns, Fannie Mae and Freddie Mac, Citigroup, and AIG to be bailed out (repeatedly), wars to be fought. . . .

Peak Coal—Richard Heinberg, Ph.D., is a world-renowned energy expert who has authored many books on the energy crisis, including one called *Peak Everything*. He has looked at coal data from around the world and come up with some startling conclusions:

> Coal provides over a quarter of the world's primary energy needs and generates 40 per cent of the world's electricity. Two thirds of global steel production depends on coal.
>
> Global consumption of coal is growing faster than that of oil or natural gas—a reverse of the situation in earlier decades. From 2000 to 2005, coal extraction expanded at an average of 4.8 per cent per year compared to 1.6 per cent per year for oil: although world natural gas consumption had been racing ahead in past years, in 2005 it actually fell slightly.
>
> Looking to the future, many analysts who are concerned about emerging supply constraints for oil and gas foresee a compensating shift to lower-quality fuels. Coal can be converted to a gaseous or liquid fuel, and coal gasification and coal-to-liquids plants are being constructed at record rates.
>
> This expanded use of coal is worrisome to advocates of policies to protect the global climate, some of whom place great hopes in new (mostly untested) technologies to capture and sequester carbon from coal gasification. With or without such technologies, there will almost certainly be more coal in our near future.

According to the widely accepted view, at current production levels proven coal reserves will last 155 years (this according to the World Coal Institute). The US Department of Energy (USDoE) projects annual global coal consumption to grow 2.5 per cent a year through 2030, by which time world consumption will be nearly double that of today.

A startling report: less than we thought!

However, future scenarios for global coal consumption are cast into doubt by two recent European studies on world coal supplies. The first, Coal: Resources and Future Production (PDF 630KB), published on April 5 by the Energy Watch Group, which reports to the German Parliament, found that global coal production could peak in as few as 15 years. This astonishing conclusion was based on a careful analysis of recent reserves revisions for several nations.[8]

The world is also shifting to coal as other energy sources become harder to obtain or become more expensive. This is posing several problems. The easiest one first: Coal is 50 to 200 percent heavier than oil per energy unit. This makes it much more expensive and energy-intensive to transport than oil. America's railway system has long been neglected, and this is the only real way to ship coal over long distances. As of March 2006 plants were running short of coal, not because it wasn't available, but because the railways couldn't deliver all that was needed.[9] If the U.S. is to rely on more coal, then the railways must be rebuilt and expanded. This is essential for several reasons, not least of which is that trains are about the most efficient form of carbon-powered transportation there is.

Then there is the final rub, proving that overpopulation and growth-fed demand are the real culprits. While coal industry lobbyists cheer that there is enough coal to last for a hundred-plus years, increasing demand and switching to coal as a replacement have changed the picture. Some energy experts are also lamenting that quoted reserve figures are about as reliable as those for oil. If transparent reserve numbers were known, we might see that Peak Coal will arrive (or has arrived) much sooner than expected.[10]

At the annual conference of the Association for the Study of Peak Oil—USA (ASPO-USA) in September 2008, board member Morey Wolfson announced that "China will build 500 coal-fired plants in the next decade." These will not be so-called clean coal plants, and the environmental risks are obvious and compelling enough without even considering what global coal depletion rates might be. Coal-fired plants are expensive and require many years of operation to provide a return on investment. After the coal runs out, then what? Will there be enough money to build some new kind of plant, running on some new kind of energy?

Forgot to tell you—Here's something all those clean-coal rah-rah commercials neglect to tell you. At the same ASPO-USA conference, University of Texas at Austin professor and associate director of its Center for International Energy & Environmental Policy said flatly, "Coal-fired power plants are the number-one user of fresh water in the country. They require 21 gallons of fresh water for each Kilowatt hour (kWh) of energy produced."

According to the Department of Energy, in 2001 U.S. residential energy consumption equaled 1,140 billion kWh.[11] That means that 23,940,000,000,000 gallons of fresh water were consumed to power our homes seven years ago. That does not include commercial and industrial electricity consumption. It also doesn't include the exponential growth that has been occurring since.

The icing on the mountaintop—Coal mining destroys open land like no other energy activity except perhaps Tar Sands oil production. It involves the removal of entire mountaintops and all the forests on the mountains themselves. It destroys entire regions and fills rivers and streams with tons of toxic effluents. According to sustainability advocate and energy researcher Mark Robinowitz, coal mining "does not take coal out of the mountain, it takes the mountaintop off of the coal."[12]

Fischer-Tropsch and Coal-to-Liquids—According to the World Coal Institute, there are two different methods for converting coal to liquid fuels:

- **Direct liquefaction** works by dissolving the coal in a solvent at high temperature and pressure. This process is highly efficient, but the liquid products require further refining to achieve high grade fuel characteristics.
- **Indirect liquefaction** gasifies the coal to form a "syngas" (a mixture of hydrogen and carbon monoxide). The syngas is then condensed over a catalyst—the "Fischer-Tropsch" process—to produce high quality, ultra-clean products.

Looks like a lot of energy input and an enormous amount of pollution, even if carbon sequestration were ever implemented. The Fischer-Tropsch process was developed in Germany and used by the Nazis as a fuel source during the Second World War. As with early ethanol, it was a net-energy loser. It failed because it took more energy to produce the diesel and gasoline from coal than what was obtained from burning it. But it was a war. Today the process is somewhat improved, and South Africa has been relying on it for some time.

Who are serious proponents of this plan? Democrats, according to the *Washington Post*.[13]

Here's a homework assignment for you. Take a few clicks on a search engine and see what it costs to build one coal-fired generating plant; then a coal-to-liquids plant. America is a participatory democracy (republic) where citizens have the responsibility to be informed. Start thinking with and as your president. It will benefit everyone. What he or she will need most is for you to understand the problems that must be solved.

Tar Sands

Tar sands are the substance, officially called bitumen, from which synthetic oil is produced. (Tar sands aren't really tar; they are just horribly sticky and thick like tar.) What the "mining" produces is not oil, but something that requires more energy to be converted into oil. So why do *CNN* and *Fox* and the *New York Times* call it oil, perpetuating in the public mind the horribly misleading impression that the problems of getting and using it are similar? Even the oil companies refer to it as synthetic oil.

Most people are familiar with Venezuela's so-called "extra-heavy oil." Just recently Venezuela reclassified all their bitumen as oil, thus allowing them to claim the largest "oil" reserves in the world. It's the same stuff as found in Canada, only closer to the surface and warmer so it flows a little more easily.

The media had a field day selling that false concept as everyone breathed an unfounded sigh of relief and went back to running up their credit cards.

Canada's tar sands are in central and northern Alberta province where it is much colder. This changes the net-energy picture drastically. The bitumen is located much deeper underground, and the process of strip-mining the tar sands is projected to destroy more than a million acres of pristine boreal (northern) forest, sometimes called the lungs of North America. That's about the size of Florida. This has brought Canada's tar sands production up to around two million barrels a day, almost all of which is exported to the U.S., making Canada America's number-one oil supplier (roughly 20% of our daily imports).

SOURCE: WWW.TREEHUGGER.COM/.../12/CANADIAN_OIL_AT.PHP.   SOURCE: GLENDONROOTSSHOOTS.WORDPRESS.COM/.../

The U.S. has announced that it expects to buy 3.5 million barrels per day (Mbpd) of tar sands oil by 2015. As of 2008 the United States was importing around 1.9 Mbpd from Canada.[14] Of what we get from Canada today, more than half is from tar sands, as Canada's conventional oil output is also in decline. On its face this sounds like a good way to start weaning the U.S. off of "foreign" oil, but it assumes that Canada belongs to the United States to exploit at will. But remember, with a global decline rate in conventional oil production of 9% or more per year, adding roughly 2 Mbpd from tar sands by 2015 (seven years after this book is written) means an increase of only about 400,000 barrels per day/year in the face of roughly 4 Mbpd/year in total annual global decline—if it's even possible. With the U.S. consuming 25% of the world's energy output and U.S. production in permanent decline since 1970 (which won't be reversed by any kind of Sarah Palin-endorsed drilling), then all we need to do is to take America's share of the decline (25% x 4,000,000) and see that we are trying to offset a loss of 1 Mbpd/year with an increase of 400 thousand barrels per day/year. We're still losing a net 600 thousand barrels per day/year!

There's that dagblasted simple arithmetic again.

The process of obtaining bitumen is fairly straightforward. You just remove millions of tons of tar sands (and clay) by stripping away the forests and wildlife on top of the land, then strip mining the top five, six or seven hundred feet of sands where the bitumen is mixed into a heavy, nearly solid paste. Then you use the world's largest dump trucks to move millions of tons to a collection area where you boil hundreds of millions of gallons of water by burning natural gas. Then you mix in caustic soda and high-pressure wash the bitumen out once it is warm enough to flow.

Processing tar sands uses enough natural gas in one day to heat three million homes, and 90% of the waste water ends up in toxic tailing ponds. Producing one barrel of "oil" produces three times the greenhouse gas of conventional oil production.[15]

And you still don't have oil yet. You have bitumen.[16] But the bitumen is still too thick to flow through a pipeline to a special kind of refinery where it can be turned into synthetic oil. You must then add either lighter petroleum or other chemicals to get the bitumen to the place where you add more energy and chemicals to refine it into gasoline.

It takes two tons of tar sands to produce one barrel (42 gallons) of synthetic oil.[17] To produce 20 Mbpd of oil, you would move 40 million tons of sand per day. Or 14.6 billion tons of sand per year.

All of this is before we get to the costs of what is the most environmentally destructive, greenhouse-gas emitting, natural gas burning, fresh-water wasting, nonsensical alternative energy source currently operating on any scale.

Shale Oil

No one claims they can actually make gasoline from this source. It has never been done on a commercial (profitable) basis. While CNN glibly tosses out the "fact" that there are "two trillion barrels of oil" in shale, they neglect to mention that it is not oil at all but another so-called "oil-equivalent" known as kerogen.

Here's the theoretical process to turn shale into oil:

> Production of oil from oil shale has been attempted at various times for nearly 100 years. So far, no venture has proved successful on a significantly large scale (Youngquist, 1998b). One problem is that there is no oil in oil shale. It is a material called kerogen. The shale has to be mined, transported, heated to about 4500C (8500F), and have hydrogen added to the product to make it flow. The shale pops like popcorn when heated so the resulting volume of shale after the kerogen is taken out is larger than when it was first mined. The waste disposal problem is large. Net energy recovery is low at best. It also takes several barrels of water to produce one barrel of oil. The largest shale oil deposits in the world are in the Colorado Plateau, a markedly water poor region. So far shale oil is, as the saying goes: "The

fuel of the future and always will be." Fleay (1995) states: "Shale oil is like a mirage that retreats as it is approached." Shale oil will not replace oil.[18]

A recent Op-ed in the *Washington Post* had a good presidential grasp on the issue. That's a good thing because the author, Ken Salazar, a t the time a United States Senator from Colorado. Here are some excerpts from his article entitled, "Heedless Rush to Oil Shale."

To hear Bush touting Western oil shale as the answer to $4 per gallon gasoline, as he did again yesterday in the Rose Garden, you would think it was 1908 . . . or 1920 . . . or 1945 . . . or 1974. Every couple of decades over the past century, the immense reserves of the oily rock under Colorado and Utah reemerge as the great hope for our energy future.

Bush and his fellow oil shale boosters claim that if only Western communities would stand aside, energy companies could begin extracting more than 500 billion barrels of recoverable oil from domestic shale deposits. If only the federal government immediately offered even more public lands for development, the technology to extract oil from rock would suddenly ripen, oil supplies would rise and gas prices would fall.

If only . . .

Furthermore, energy companies are still years away—2015 at the earliest—from knowing whether this technology can cost-effectively produce oil on a commercial scale . . .

. . . It would take around one ton of rock to produce enough fuel to last the average car two weeks.

. . . How is a federal agency to establish regulations, lease land and then manage oil shale development without knowing whether the technology is commercially viable, how much water the technology would need (no small question in the arid West), how much carbon would be emitted, the source of the electricity to power the projects, or what the effects would be on Western landscapes? . . .

The governors of Wyoming and Colorado, communities and editorial boards across the West agree that the administration's headlong rush is a terrible idea . . . [19]

In the decidedly mixed-bag policies and appointments of President Barack Obama the appointment of Salazar as secretary of the interior is perhaps the brightest point of awareness.

If it would take one ton of rocks to power one car for two weeks, how many tons of rock would it take to power 270 million cars for one year? Answer: 7.02 *trillion* tons. Now add the hydrogen (I'm tired of talking about it) and probably the equivalent all of the fresh water flowing in the Colorado River every year and you might have something. Oops, we forgot about the infrastructure costs didn't we?

We forgot about irrigating crop lands and drinking water. We forgot about the hydroelectric power generated at Hoover Dam.

How much energy is used to heat shale to one third the temperature of the sun?

Not surprisingly, just before leaving office the Bush administration relaxed or suspended EPA regulations to benefit shale oil promoters and developers. The Obama administration under Salazar's guidance thankfully reversed those orders quickly.

Nuclear

Over the years I had saved close to 800 pages of stories about nuclear energy. In updating that research for this book I went through another 80–120. I realized that there was a real danger of turning a short book into an encyclopedia because there are so many new types of reactor technologies being proposed and "on the drawing boards." They are all theoretical and as such by definition are of little practical relevance now. But as I separated the pro and con stories into two piles and kept going through them

I saw some underlying truths that (while sometimes ignored) are not in dispute from either side.

- Currently there are just over 100 nuclear power plants operating in the United States. No new plants have been built in more than 30 years.
- There is not enough uranium to build enough reactors to meet energy demand globally. "Current global uranium production meets only 58 percent of demand, with the shortfall made up from rapidly shrinking stockpiles. The shortfall is expected to run at 51 million pounds a year on average from next year to 2020.[20]
- Reactors cost between 2.5 and 6 billion dollars per plant and take five to seven years to build AFTER the two-to-four year permitting and licensing process is complete. This is called the gestation period.
- Nuclear power cannot and will not displace the use of oil (which mostly powers transportation).
- Enormous amounts of raw materials and especially hydrocarbon energy and electricity are required to enrich uranium into a useable energy source. Uranium in the ground is useless. The process of enrichment—as we have all seen in recent years with Iran—takes enormous amounts of energy and capital (i.e. energy investment). The building of nuclear generating stations produces enormous amounts of greenhouse gas and other pollutants. The process of enriching uranium also creates enormous amounts of greenhouse gas and other toxic waste.
- Nuclear plants require enormous amounts of security infrastructure that can only be provided or overseen by a national government.
- Nuclear power plants pose a risk of devastating malfunctions as witnessed by the Three Mile Island and Chernobyl disasters.
- Nuclear plants produce toxic waste that is the most lethal

Expected New Nuclear Power Plant Applications Updated September 28, 2009						
Company*	Date of Application	Design	Date Accepted	Site Under Consideration	State	Existing Operating Plant
Calendar Year (CY) 2007 Applications						
NRG Energy (52-012/013)***	09/20/2007	ABWR	11/29/2007	South Texas Project (2 units)	TX	Y
NuStart Energy (52-014/015)***	10/30/2007	AP1000	01/18/2008	Bellefonte (2 units)	AL	N
UNISTAR (52-016)***	07/13/2007 (Envir.) 03/13/2008 (Safety)	EPR	01/25/2008 06/03/2008	Calvert Cliffs (1 unit)	MD	Y
Dominion (52-017)***	11/27/2007	ESBWR	01/28/2008	North Anna (1 unit)	VA	Y
Duke (52-018/019)***	12/13/2007	AP1000	02/25/2008	William Lee Nuclear Station (2 units)	SC	N
2007 TOTAL NUMBER OF APPLICATIONS = 5 TOTAL NUMBER OF UNITS = 8						
Calendar Year (CY) 2008 Applications						
Progress Energy (52-022/023)***	02/19/2008	AP1000	04/17/2008	Harris (2 units)	NC	Y
NuStart Energy (52-024)***	02/27/2008	ESBWR	04/17/2008	Grand Gulf (1 units)	MS	Y
Southern Nuclear Operating Co. (52-025/026)***	03/31/2008	AP1000	05/30/2008	Vogtle (2 units)	GA	Y
South Carolina Electric & Gas (52-027/028)***	03/31/2008	AP1000	07/31/2008	Summer (2 units)	SC	Y
Progress Energy (52-029/030)***	07/30/2008	AP1000	10/06/2008	Levy County (2 units)	FL	N
Detroit Edison (52-033)***	09/18/2008	ESBWR	11/25/2008	Fermi (1 unit)	MI	Y
Luminant Power (52-034/035)***	09/19/2008	USAPWR	12/2/2008	Comanche Peak (2 units)	TX	Y
Entergy (52-036)***	09/25/2008	ESBWR	12/4/2008	River Bend (1 unit)	LA	Y
AmerenUE (52-037)***	07/24/2008	EPR	12/12/2008	Callaway (1 unit)	MO	Y
UNISTAR (52-038)***	09/30/2008	EPR	12/12/2008	Nine Mile Point (1 unit)	NY	Y
PPL Generation (52-039)***	10/10/2008	EPR	12/19/2008	Bell Bend (1 unit)	PA	Y
2008 TOTAL NUMBER OF APPLICATIONS = 11 TOTAL NUMBER OF UNITS = 16						
Calendar Year (CY) 2009 Applications						
Florida Power and Light (763)***	6/30/2009	AP1000	09/04/2009	Turkey Point (2 units)	FL	Y
Amarillo Power (752)		EPR	Vicinity of Amarillo (2 units)	TX	UNK	
Alternate Energy Holdings (765)		EPR	Hammett (1 unit)	ID	N	
2009 TOTAL NUMBER OF APPLICATIONS = 3 TOTAL NUMBER OF UNITS = 5						
Calendar Year (CY) 2010 Applications						
Blue Castle Project		TBD	Utah	UT	N	
Unannounced	TBD	TBD	TBD	UNK		
2010 TOTAL NUMBER OF APPLICATIONS = 2 TOTAL NUMBER OF UNITS = 2						
Calendar Year (CY) 2011 Applications						
No Letters of Intent have been received from applicants expressing their plans to submit new COL applications in CY 2011.						
2007 – 2011 Total Number of Applications = 21 Total Number of Units = 31						
* Project Numbers/Docket Numbers ** Acceptance Review Ongoing *** Accepted/Docketed						

New nuclear construction is beset with many problems including heavy upfront capital investment and very long lead times to ensure safety and security. It might take a decade to bring a new plant online and would not make a difference in immediate energy shortages. SOURCE: HTTP://WWW.NRC.GOV/REACTORS/NEW-REACTORS/NEW-LICENSING-FILES/EXPECTED-NEW-RX-APPLICATIONS.PDF.

substance known to man, and it stays that way for hundreds of thousands of years. The amount of energy required to imprison and monitor the wastes for thousands of years is incalculable. It is supremely selfish to pass these unknowable costs to generations that may not have the energy, money or technical ability to complete the task.

In 2007 only five applications were received by the Nuclear Regulatory Commission for a total of eight new units. The NRC preview process can take years before permits are issued to begin the five- to ten-year construction process. In July 2008 the number had risen to thirteen applications for nineteen units, but no new plants had even been approved and no ground had been broken even though serious energy shortages have been with us for at least six years.

Have you heard of some other fantastic new alternative energy that I did not mention here? All you have to do is to go back through this and the preceding chapter and evaluate it for yourself. Now you know what to look for. I refuse to do that work for you. It is not energy-efficient.

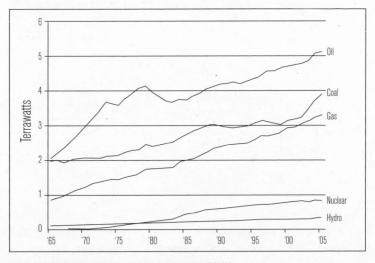

SOURCE: HTTP://EN.WIKIPEDIA.ORG/WIKI/IMAGE:WORLD_ENERGY_CONSUMPTION.PNG

THE PRESIDENT'S NOTES

As the president evaluated all of these options he or she might have been making notes that look something like this:

- We're screwed—Situation worse than thought. We need alternatives to the alternatives! Alternatives either limited or smoke and mirrors. Pursuing some looks like (and is) desperation to partially replace oil and gas—not confident, actual switching. Ransacking planet and lives in search for energy.
- Renewables cannot support the edifice built by fossil fuels.
- Clear tradeoff bet. energy and environment. Tip all the way one way and society breaks down for lack of energy, jobs + crime, poverty & dislocation . . . —tip full opposite, kill planet and ourselves. Must find best balance in order to transition.
- Natural gas precious. Watch shale gas closely. NG clean-burning and efficient. Most important use electrical generation. No full picture on how damaging shale gas is. Where is EPA? Looks real bad. No cost (net energy) data. Shale gas could gain votes nationally in the short term but loose votes of ranchers, farmers and those areas polluted or sucked dry of ground water (e.g. New York City). What about land damage and food? Robbing Peter to pay Paul?
- The World Nuclear Association predicts a 40% jump in global demand for electricity between now and 2012.
- Push solar and wind all out. Need to look at what geographic areas of U.S. don't or can't benefit. Same with tidal. Which regions are most at risk

then? Need to ident. regions/ help assess their options on priority basis. What is fedgov role? What is local?

- Nuclear increase may be inevitable but very limited and very dangerous. Not enough time to build plants to offset declines and growth. Not enough uranium. Not enough money. Maybe not enough time to build before climate for safe building fades. What the heck do we do with the waste?

- Coal increase inevitable but what about climate change? What about running low on coal as we use to fill other needs? What about environment?

- Tar sands—Canadian environmental issue but caused by U.S. demand. Canada short of nat. gas. Must continue to limit U.S. gasoline demand. Continued use of tar sands inevitable but no solution. Probably won't ever exceed 3–4 mbpd due to cost, nat. gas/water shortages and things falling apart. Global decline will outstrip that in a heartbeat

- Pressure on news editors, corporations and all media to stop promoting oil shale, hydrogen and coal-to-oil. Only making my job harder. Only making it harder for citizens to understand. Must exert pressure to stop selling ad and commercials to entities that lie or mislead. Bonehead sales pitches make my job harder.

- How can fedgov incentivize the technologies that might help?

- Must use less and less oil and gas. Save what's available in face of decline for most important uses—to hold things together.

- Need to educate the people. Must educate Congress. Where's the sledgehammer to hit the Ox between the eyes.

CHAPTER ELEVEN

Localization: The Alternative To The Alternatives

The end of the Age of Oil will also be the end of globalization[1], long-distance commutes, and long-distance transportation of goods and services—period. Oil is the *only* transportation fuel we have today, and it will be for some time. As president, you grasp that there is not going to be a last-minute reprieve from some new magical solution, a secret weapon that is going to win the "war" at the last minute. You look around and realize that localities are bearing the brunt of the hardship and you ask yourself what your role—what the role of the federal government—should be.

Since most Americans live in or near large cities, it might be best to hear what the cities are saying themselves. For almost every city in America oil is the single largest budget item, and in 2005 Denver's oil costs surged by $1.9 billion. In 2005 and 2006 *From the Wilderness* attended conferences of the Association for the Study of Peak Oil, USA (ASPO-USA) in Denver and Boston. In Denver, Mayor John Hickenlooper made it clear that cities were bearing the heaviest burden because it was cities that delivered the services that mattered most to people.

Ad hoc networking had begun between many cities around the country to share ideas on efficiency, conservation and alternatives. Denver was sharing information with Portland and Chicago. Emergency energy task forces were sprouting up everywhere. Peak Oil was not speculation for these folks but a given at the local level, and there was serious frustration with the federal government, which was perceived as being "out of touch" as unfunded mandates on climate change and greenhouse-gas emissions strained municipal treasuries. Costs were being pushed down from Washington.

By 2008 the gap between what cities needed and the federal government was doing had worsened. I made several contacts with lobbying groups dealing with municipal issues in Washington. All sources spoke on a not-for-attribution basis but were very clear in their positions. "The federal government just doesn't get it," was said by more than one source. The general consensus was that by imposing unfunded mandates on climate issues and by continuing to build new roads through cities or major interstates, even as traffic flows were shrinking, the federal government had become a "huge drag on cities' ability to respond to rising fuel costs and what that does to other services in municipal budgets." The cities are now crying for what they call "reverse mandates" where cities can tell the federal government what is needed in the way of block grants that could be applied by local governments.

Some of the language was strong. "We're getting creamed in every direction. The costs of capital improvements are like double-digit inflation. The federal government uses its resources in the most reckless and inappropriate ways."

As president, your first awareness is that the federal government cannot and will not take on the role of solving problems in cities and townships. That would be inefficient and inappropriate on every conceivable level. Only the people in each locality know and can decide what they need most. Each location has different needs.

Your second awareness is that if localities fail at the bottom, the nation will fail at the top. Tax revenues are shrinking at every level of government. Federal employees all over the country are already having trouble getting to work because of economic challenges. That problem is going to worsen. America's "all-volunteer" military will shrink because sons and daughters will have to stay at home to help support increasingly distressed families.

Decline is a fact that is not going to go away even if a million wells are drilled. Drilling holes does not mean that oil will be there. We might have better luck in Las Vegas or Atlantic City, although it's pretty clear that these cities have short life expectancies. As oil supply tightens, the ability of the nation and of each community in it to respond effectively to problems, or to simply function at all, will be dictated by its degree of self-sufficiency and the degree to which it has liberated itself from dependence on anything from outside, whether the outside is 150 miles away or across an ocean. Food is the first concern here. Somehow America must start producing food where it is eaten, the way it did in the 19th century.

Awareness of this truth and some preparations for it have been underway for several years, slowly at first, but now at an accelerating pace—always at the local level, frequently at the individual household level. Yet many long-term activists, organizers and planners in the field complain that such preparations are far behind where they need to be.

As president, you might sit down one morning at your desk and receive the following briefing.

- The Montana State Highway Patrol is cutting back on law enforcement patrols because it cannot afford the cost of gasoline.[2] In fact almost every state highway patrol has been reducing services which has increased response times to emergency situations. Local police departments are faring no better. The El Paso County Sheriff's Department in Colorado has ended car patrols within its 2,000-square-mile jurisdiction. One Ohio sheriff is putting his deputies into golf carts. Stillwater, Okla., has stopped mowing the grass on nearly half of its parkland.[3] That seems like a small problem but it isn't when untended parks provide havens for criminals and homeless. The Attorney General and the FBI are advising you in confidential briefings that crime rates in rural areas are rising rapidly as a result of unemployment and a growing awareness that police patrols have been cut back.

- School districts around the nation have begun

cutting back to 4-day school weeks because they can no longer afford the fuel for school buses.[4] This is also causing an increase in juvenile crime and resulting in less-capable graduates entering an already overloaded work force

- Asphalt prices for road maintenance have risen so high that roads are starting to disintegrate, causing damage to private vehicles that fewer venues can afford to repair. A new report from Maryland indicates that asphalt prices have made it impossible to repair a road to a local church. A report from a local newspaper reports a discovery that every city and county in America is making, "To put it another way, a 20-foot wide, mile-long road with 2 inches of blacktop cost the county about $51,000 in 2004. That same road now costs $98,213."[5]

- In fact, 90% of U.S. cities are cutting back services because of fuel costs, everything from police and fire to trash pickup has been hit. "Several mayors—as they gripped-and-grinned at a downtown hotel—said the cost of fuel had become their obsession."[6] If trash isn't picked up then disease becomes a risk, and public health becomes an issue. Larger cities have not been forced to cut back on emergency services yet, opting instead to cut back on lower priority social services and discretionary programs as rising energy costs sap their budgets. That can only last for so long.

- All over the country, shelves at local food banks are empty, as food and transportation costs have collided headlong with a collapsing economy.

- Displaced populations suffering from home foreclosures and unemployment are colliding with a new exodus moving into inner cities where those able to afford it have decided to move to cut commuting costs.

The list seems endless and is growing longer every day. All of these conditions are only magnified by the current depression.

Advisers have suggested to you that the federal government might start issuing mandates on energy use and setting standards for continued receipt of federal aid. As to the second point, you have decided that some standards might be necessary, but you have ruled out ones that require cities, counties, and states to spend money they no longer have. California's 2008–9 budget crisis has made it clear that unfunded mandates might only accelerate the breakdown instead of slowing it.

One bright spot is that many thousands of individuals and families have, on their own, moved to make themselves less dependent upon fossil fuels and outside goods and services. A problem here is that there is no data base to track these efforts or what has been learned that may be exportable to other areas. There is no clearing house for data on individual initiatives producing a solution in Wyoming that could be applied to Massachusetts or Georgia.

The Perfect Laboratories

You look around to see if any place else has had to cope with sudden and severe oil shortages. Fortunately there are two clear and unequivocal examples. One shows what works and the other reveals what doesn't. They are Cuba and North Korea, both of which experienced a sudden and dramatic absence of oil and natural gas products after the collapse of the Soviet Union in 1991. Neither country had any significant domestic energy sources although North Korea does have some coal. Both were totally reliant on oil and fertilizer (made from natural gas) exports from the Soviet Union. Agriculture in both countries had become dependent upon petrochemicals as we saw in Chapter Eight. But rather than a serious decline in the availability of oil and gas, for both of these nations it was almost an instant cold-turkey withdrawal.

One nation, following a rigid Soviet-style, top-down management system starved and ultimately nearly failed. Its populace suffered horribly as a complex civilization collapsed on all fronts.

Trains didn't run. There were massive blackouts. Frequently there was no water pressure. There was no fertilizer. That nation was North Korea.

The other nation turned almost immediately to local entrepreneurial capitalism and private ownership. It not only survived but ultimately became much healthier after a serious period of hardship. Its government made land ownership available to anyone who would farm it, even taking fallow land away from landowners who were not using it. It mandated local food production because not only was there no gasoline to drive food around the country there was almost none to power tractors and harvesters. There was no electricity to power irrigation pumps. It lifted all government interference and let the free markets operate in a way that would have made Adam Smith proud. The nation that survived was Cuba.

Cuba's transition was by no means easy. Its soil had been harmed by decades of dependence on ammonia-based (natural gas) fertilizers and monocropping. So the first and immediate task was soil restoration. During that time only a few crops were grown. But, as time passed, the Cuban diet expanded from basic subsistence to become healthier and more diverse than ever before. Not only that, rooftops and vacant lots, almost every available square inch of land in Havana became local farms and markets within easy walking distance. Barter replaced cash. All food production became organic. Large state-owned farms were broken up, much as large American "agribiz" farming operations will eventually be broken up, out of necessity. Entropy makes everything break down into smaller components.

Notwithstanding the much better climate in Cuba and a series of natural disasters that hurt North Korea's agricultural base, North Korea did everything wrong. The national government took strict control of almost every aspect of food cultivation. Cuba liberated it. After a few years of hardship Cuba's population became healthier, the diet diversified and food choices increased dramatically. American film makers travelled to Cuba and documented this inspiring transition, showing that it is possible to survive and eat well after a loss of oil and gas.[7] What proved essential was not for the national government to take control, but rather to get itself out of the way. Hunger drove the population to change its thinking or starve.

In 2003, my newsletter *From The Wilderness* published a two-part series by the brilliant Dale Allen Pfeiffer who had written our earlier story "Eating Fossil Fuels." Titled "Drawing Lessons From Experience" the series contrasted the experience from both countries. Near the end of Part II Pfeiffer wrote:

> The World Bank has reported that Cuba is leading nearly every other developing nation in human development performance. Because Cuba's agricultural model goes against the grain of orthodox economic thought, the World Bank has called Cuba the "anti-model." Senior World Bank officials have even suggested that other developing countries should take a closer look at Cuba.[43] This despite that fact that the Cuban model flies in the face of the neoliberal reforms prescribed by both the World Bank and the IMF.[8]

Megan Quinn-Bachman is the Outreach Director of the Arthur Morgan Institute for Community Solutions of Yellow Springs, Ohio—one of the most active relocalization organizations in the world. She is also the co-producer of *The Power of Community: How Cuba Survived Peak Oil*, the 2005 award-winning documentary demonstrating that relocalization is the most effective way to deal with energy limitations. She has travelled to many countries and lectured all over the U.S. and Canada on the subject. In her late twenties, well-educated and an engaging speaker, she promises to be an important future leader for the generations that will have to deal with the worst parts of the energy/food crisis. I contacted her and asked her what information she had on relocalization efforts around the country and how such efforts were progressing. Her answer was less encouraging than I had hoped.

"Well . . . there are some localities taking steps. The municipalities showcase some bits and pieces of a good strategy (peak oil resolutions, planning & zoning that preserves land and resources, energy conservation for municipal energy use, etc.) but no place has put it all together and no place will survive the crisis without major challenges. Five years ago it was lip service, now its tokenism and a piecemeal approach. No community I know of has made community-wide (not just municipal buildings, fleets, etc.) energy reduction mandates, nor has any fully embraced a transition towards import substituting businesses and local living/security. None that I know of are storing emergency liquid fuel supplies and setting up emergency warm spaces and food preservation and storage facilities. None are figuring out how to prevent resources and money from leaving the community via banking and purchasing into local business incubators for entrepreneurs to set up local, low-energy businesses and infrastructure improvements (retro-fitting homes, sustainable wastewater management, etc.).

Much of the best work is happening outside of municipalities, via community groups, neighborhoods, businesses, and pioneering individuals. CSAs [Community Supported Agriculture], food co-ops, local currencies & trading systems, are a few.

Relocalization groups, Transition Towns (http://www.transitiontowns.org/), etc. are all good starts, but they haven't gone much beyond community education. They could be a great structure to use for disseminating viable options and models, but that's currently not their use."

Perhaps the one municipality in the United States with the biggest head start on relocalization is Willits, California.

A simple truth is all too apparent. There is no hope for any of us outside of a community. We must learn to work with our neighbors in developing sustainable lifestyles based upon reduced consumption and sharing of resources. This is difficult for Americans brought up on rugged individualism and competition and who have been taught to measure success in terms of consumer goods possessed and energy expended. But this is how our ancestors, the first settlers of this country, were able to survive and thrive. It is also how the Native Americans before them survived in a sustainable balance with the land and nature. Are we so deluded as to believe there can be no joy in life without rampant consumption?

A wise man once said that success was not having what you wanted but wanting what you had. Perhaps through relocalization, if it is embraced before it becomes an imperative, we will rediscover a quality of life that we have been missing and fill the void that we have been attempting to fill with consumption.

Either way, relocalization is going to happen. We can go there by choice, or we can resist and let our children suffer for our lack of vision. Some of the great champions of Peak Oil and sustainability like Jason Bradford and Matt Savinar live there. The web site http://www.willitseconomiclocalization.org/ is—as far as I can tell—a cutting edge of relocalization planning and experience.

CHAPTER TWELVE

Money

Here we arrive at the real heart of the issue. It is the way that money works that has locked us into an inevitable collision of two mutually exclusive operating principles. A manmade requirement for infinite growth collides with a man-sustaining and unyielding finite planet and the physical laws that govern our universe. For the religious (or spiritual) reader I have no quarrel with God at all. When this planet was created, it was given certain laws to govern it, which are definitely "bigger" than mankind. Any way you define it, that *is* God, because these laws are more powerful than we are as humans. Scientific laws do not change because of mankind's influence—or demands. Gravity will always be gravity and the Laws of Thermodynamics will not reverse because we might wish it so.

Previously, we looked briefly at how the world's economic paradigm is currently operating, but it is well worth looking at it in just a little more detail; in a way that won't make the reader's eyes glaze and brain go limp. This is actually very straightforward. The only thing that needs to be opened to grasp it is your mind.

Fiat Currency

The world makes money by printing it. Money no longer has any connection to anything tangible. In ancient civilizations, and up until the modern day, money was always based upon something that was physical, e.g. gold or silver, food or trade goods. The British Pound Sterling was a classic example of this. It was linked to one pound of silver. There was only a certain amount of that precious metal in existence and, historically, money was pegged to that tangible substance, either in whole or in part. By saying that there could be no more than x number of pounds or dollars in relation to a nation's store of silver or gold, there was a naturally self-imposed limit on how much money could be in circulation. Money was used to open businesses and to buy or make things that other people then bought. This connection to something tangible provided for very stable currencies, but limited how much growth could take place by connecting it to the physical limitations of the earth.

Over the course of the last 150 years, every major nation in the world has steadily decoupled their currencies from this fundamental fact.

The United States started separating itself from the gold standard with the creation of the Federal Reserve System in 1913. It finished the process in the Nixon administration in the 1970s when all remaining legal requirements linking dollars to gold reserves were removed. U.S. silver and gold certificates disappeared and all U.S. currency became Federal Reserve Notes.

The Federal Reserve (or Fed), by the way, is not a U.S. government institution. It is, as some have aptly noted, no more federal than Federal Express. It is a consortium of privately owned banks which can print whatever amount of money it deems necessary, without regard to physical limitations. Its basic rule is: print too much money and you have inflation, which is bad; print too little and there is no growth, i.e. a recession or a depression, which is also bad. The "bad" is only "bad" as it applies to the monetary system of endless growth which we are examining. It has nothing inherently to do with the well-being of people, nations, or of the planet. We have been led to believe that taking care of money was the same as taking care of ourselves when, in fact, the opposite is true. Taking care of money kills people.

Who actually owns the Fed? Although not revealing percentages or stakes, a 1976 congressional hearing listed a group of names which may sound all too familiar now. They include names like Rothschild, Rockefeller, Warburg, Morgan, Lehman, and Alex Brown (closely connected to the Bush family).[1]

This policy helped stimulate the unprecedented growth of the U.S. economy at the turn of the last century, as it coincided with the newfound abundance of available cheap energy and a seemingly inexhaustible natural resource base. That made possible a population explosion which added five billion new consumers to our planetary fold and more than doubled the U.S. population. As the infinite-growth economy expands, the number of consumers must continue to grow also.

No one seemed to grasp the fact that money cannot be decoupled from energy. For without energy and what it produces (e.g. food), money is valueless. You cannot eat a $100 bill, nor can you put one in your gas tank and expect to get any benefit from it.

Fiat currency is any currency that is created merely by a directive to print it—a fiat.

Energy, however, which is essential to give money value, cannot be created out of thin air. Presidential candidates and presidents cannot issue a directive creating more energy. The First Law of Thermodynamics overrules them. Energy can neither be created nor destroyed. The Second Law of Thermodynamics locks them into a dilemma they cannot escape from: energy only converts in one direction, from useable to unusable, and some of the useful energy is always lost in the transaction—entropy. Things always break down into smaller and not larger units.

Fractional Reserve Banking

This is very easy to grasp. Although made possible by a fiat currency, fractional reserve banking says that any bank may create money all by itself based upon the number of actual dollars it has on deposit. Current U.S. banking regulations vary, but a safe example is to say that if the ABC Bank has $100 on deposit it can make loans of about $900 in keeping with its reserve requirements as established by the Fed. Although reserve requirements vary, the rule of 9:1 is a good example. This is more money out of thin air.

The essence of the great Wall Street meltdown of September

and October 2008 resides in the fact that with all the bad and fraudulent debts from the housing bubble on their books, banks have lent all the money they can without exceeding their reserve requirements. That's why credit dried up. The fastest suicide for the financial markets would be for the Fed to lift reserve require- ments, because that would touch off the all-consuming forest fire of inflation. I believe that hyper-inflation is inevitable, especially since—according to its own data—the Fed has been printing money at a rate 100 times higher than during World War II since the third quarter of 2008. In essence, the so-called bailout bill(s), which most experts always knew was/were insufficient, allowed some institutions to take away the bad loans and create more money so that they could issue more credit.

The massive banking consolidation, the wave of acquisitions happened so that bigger banks could acquire the deposits of fail- ing banks and use that cash to service their debt and lend more money.

If you go in to borrow $25,000 to buy a car, the bank does not take $25,000 out of a cash drawer. It just makes an entry in a ledger and writes a check. Sure, that $25,000 may go to Toyota as the balance after your down payment on a new Prius hybrid, but nothing tangible ever changes hands. These are all ledger entries.

Compound Interest And Debt-Based Growth

Almost all economic growth is accomplished by borrowing. Start-up companies borrow to set up production lines and infra- structure. Established companies borrow to finance everything from major new projects that would otherwise eat all of their cash to pay for or to buy other companies (leveraged buyouts). You and I (used to) borrow from our credit cards for our car loans and for our mortgages. In every case more money must be paid back than was initially borrowed. That can happen only if there is growth.

Compound (rather than simple) interest says that for that

portion of the debt that remains unpaid per month or per year more interest is charged on the unpaid balance.

Simple interest, on the other hand, fixes the amount of money created with each transaction. A simple-interest contract of 5% says that on a $1,000 loan a "fee" of $50 will be paid for using the bank's money. A compound-interest loan says that 5% of the unpaid balance will be charged each year until the loan is paid in full. If only $100 of the loan is paid in the first year, then 5% of the remaining $900 (or $45) will be charged the next year and so on until the loan reaches a zero balance. In the third year, assuming 10% of the principle is paid leaving a balance of $810 the interest will be $40.50. In just three years $135.50 of new money has been created on a $1,000 loan.

Most young Americans have learned the hard way that if all they do is to pay the minimum monthly payment on their credit cards, it will take years to get out of debt. They will also pay back much more than they borrowed. They will be creating money in this process that didn't exist before. It's a funny thing with money. Take many years to pay off a credit card and the debtor has created lots of money, which goes back to the bank, which can (needs to) lend nine times that much again to keep the whole thing from collapsing. Inflate the value of a home arbitrarily (criminally) and lock the home buyer into a 30- or 40-year adjustable rate mortgage and you start creating lots of money that, because of the economic paradigm, must get used again—to create more money.

It's a pyramid scheme. Forget Madoff and Stanford, the entire global economy and monetary system is a pyramid scheme and always has been.

This is why I laughed so hard when I heard of a Northern California millionaire who was so concerned about Peak Oil, and liked Priuses so much, that he went out and bought five of them. He was sure that would help solve things. He paid cash, but Toyota looked at the skyrocketing sales of Priuses all over and saw they were awash in "cash" and used it to finance new plants where they could build more new Priuses—with seven gallons of oil in each tire and many hundreds of gallons of oil in the plastics, paints, sealants, vinyl, rubber insulation and foam seats. Toyota didn't pay cash for the plants. No major manufacturer does that. They made a down payment and used the rest of their fiat currency for other purposes (i.e. to stimulate growth).

And the one thing that few people think of is that for all the people who buy Priuses, they each put a "new" used car on the market for someone else to buy. The whole thing works only as long as everything keeps growing. The cycle of growth is what creates bubbles—or irregular growth patterns—which take on the characteristics of a nuclear reactor at critical mass. Chernobyl demonstrated the end result of that.

I used Toyota only as an example. I could just as easily have said Ford, Citigroup, Disney (ABC), Newscorp (Fox), Time Warner (CNN), G.E. (NBC/MSNBC), or Viacom (CBS). It's all the same paradigm isn't it? That's the point. Every major media outlet (a corporation which publicly trades stock and borrows money) in the United States is also locked into this paradigm, and that's why they absolutely refuse to discuss this. They are dinosaurs of a dying paradigm, about to go extinct, and they cannot even see (and dare not acknowledge) anything inimical to their interests. The way money works dictates that they instead try to bend everything else to money's need for growth.

We might add another law of physics that says, "All bubbles burst."

Compound Growth

Here is where it gets really ugly. China and India have both been boasting economic growth rates of 8% per year. Compound growth works the same way in this direction too. If, for example, China had 300 million cars in 2004, an 8% growth rate would suggest an increase of 24 million cars in one year. But in developing nations the first thing everyone wants is a car, so vehicle growth rates are actually much higher.

Take a chessboard. Put one grain of rice on the lower left hand square. In the square to its right put two grains. Double that to four for the third square and keep doing that until you reach the last square. An Australian government web site looking at the issue of compound growth reported the following. "The number of grains was increasing as a geometric progression, and the total amount of rice required to fill a 64-squared chess board is (2^64 − 1), which is equal to 18,446,744,073,709,551,615 grains (about 18*10^18, or 18 billion billion grains). This amount of rice would weigh about 461*10^12 kg, or 461 billion [with a "b"] tonnes."[2]

The UN's Food and Agriculture Organization (FAO) predicted that the 2008 global rice harvest would be 666 million (with an "m") tons, or only 14% of what appeared on your chessboard. Compound growth totally outpaced available resources in a very short time.

In 2002 China's demand for automobiles increased by 56%. An article in Britain's *The Economist* reported, "As long as the economy goes on galloping at its current high-single-digit clip, many expect car sales to increase by 10–20% annually for several years to come . . . "[3]

On July 17, 2008, the "China Economic Blog" reported the following:

> We project that the number of cars will increase by 2.3 billion between 2005 and 2050, with an increase by 1.9 billion in emerging-market and developing countries.[4]

As I was re-reading this passage, I noticed that it was unclear whether the 2.3 billion vehicle increase was going to be in just China or the world as a whole. I started to go look it up, and then I stopped—laughing at myself really hard. It doesn't matter does it?

The UN and other sources estimate that there are *only* around 700–800 million internal-combustion-powered vehicles on the planet right now. Having looked at alternative energy sources, and seeing that no infrastructure or technology exists to make anything else, then virtually every one of these vehicles—if they are made—will be powered by gasoline. It doesn't matter if they get 200 miles per gallon. Growth will outpace available resources.

Of course all these vehicles will never be built. Ever. Certainly not with seven gallons of oil in every tire. You do that math yourself. How much oil to build the factories and ship the raw materials?

So how does one tell the people in China and India, in all developing countries, "Oops, sorry. You are not going to get the American dream that we have marketed so well through our movies and pop culture. We used it all up"? The Indian press vehemently and irrationally denies Peak Oil from what I have seen. This again is the first stage of grief. The global anger that is sure to follow the realization of what I have just described has not even begun yet.

American culture and consumption has become Public Enemy Number One in the global growth paradigm. People are realizing that the American Dream is murder. I wonder when the American people will grasp that the American Dream, as they conceive it, is a marketing gimmick intended to sell consumer goods. For some reason I have held the belief that the American Dream had to do with life, liberty and the pursuit of happiness; freedom of speech, religion and assembly; the right to petition for redress of grievances. Silly me. I also never accepted advertising agencies' subliminal brainwashing that consumption was the same thing as happiness.

Collapse

Obviously, this is not sustainable. The term "sustainable growth" is the quintessential oxymoron. Those expecting to get repaid can only do so if there is a continuous stream of new borrowing (i.e. growth) at the bottom. Those at the bottom can only make payments if they grow, and they will only make money if people buy their products or services—frequently by borrowing . . . ad infinitum. Until there is no more energy.

Unless a fundamental change is made—and quickly—the only available option is implosion and collapse; the bursting of the human population bubble; or, as people in the Peak Oil movement call it—the Die Off. The sole purpose of this book (and my life) is to prevent that, or as much of that death and misery as is humanly possible.

That's why I have said for years in writing and in my lectures, "Until you change the way money works, you change nothing." M. King Hubbert, the Prophet of Peak Oil, was the true visionary. Six decades after he began looking into the subject of Peak Oil and non-renewable resources we see that he was absolutely correct when he said that the only way out of the infinite growth trap was to create a "steady-state" economy—an economy without growth.

That is much easier said than done. However, it is where the laws that God put in place will inevitably take us whether we like it or not. The harder we fight it, the greater the die-off and suffering will be . . . threatening all life on the planet. The more we move willingly in that direction the greater our chances for survival as a viable species. Our only hope for survival is a complete surrender to this inevitability.

It is here that a modern world, created and ruled by corporations more than governments, collides headlong with issues that pertain to the survival of the species. I think it fair to say that such a fundamental change in human behavior is beyond the power and wisdom of any American president. Here, all of civilization, beginning with the first cultivated crops some 30,000 years ago, collides with physical reality.

It is hard to find rational thought on the subject anywhere because this physical truth is as unsettling to entities like the Roman Catholic Church as was Galileo's notion that the earth revolved around the sun. They locked Galileo up for that, even though his navigational science based on a heliocentric solar system permitted Catholic vessels from Spain, Portugal, Italy, France, Holland and England (after the Reformation) to sail off in search of new worlds to conquer, new slaves from which to draw energy, and new resources to exploit.

The Catholic Church is by no means the only institution or icon that must be considered—although it may be the most important. If the Roman Catholic Church were to ever rethink its positions on birth control and growth, then the door would be opened for the whole world to rationally discuss the subject. There are other villains. Every population that values boy children more than its girl children; every tribal society that prizes more children as a sign of prestige or ability to hunt, gather, fight and farm must be somehow breached with a simple awareness that there are too many of us. We are harming or killing everything and committing suicide in the process. Essentially the Catholic Church's position is that the human race must kill everything—including itself—in order to make more babies.

I cannot help but note the collective anger expressed by the birth of octuplets to an artificially inseminated, unemployed, single woman in California in early 2009. Although I can pity the woman—and especially the children—I cannot blame her. I am heavily influenced by Jung and related spiritual teachings, and I recognize that in our dysfunctional world this woman has merely manifested the insanity of a position we have all endorsed and enabled as a species. Another sad part of our human nature is that we might wish to symbolically punish her in order to make ourselves feel just a little more righteous.

All of this is inextricably tied to the way money works and all of this is beyond the real reach of any American president—save for the "bully pulpit" and the unprecedented opportunity to lead by example which the office brings with it. Were an American president to initiate a serious discussion on these subjects in this country—the one which uses the most energy (and most energy per capita)—then the world would be compelled to follow suit.

There are no more worlds left to conquer and exploit. There

will not be enough cheap energy to grow more crops on ever-depleting and shrinking farmlands. This is a battle that Mother Earth is absolutely certain to win and, unless it changes its ways, mankind is absolutely certain to lose. Against this, perhaps the greatest challenge ever faced by mankind, the American people can do only little things. But some of them may have profound differences that will help to crack the intransigence of the human psyche and its unwillingness to address the only two real issues there are: the way money works and population growth.

Beware The Green Investment Bubble

There is much popular talk about the coming new Green Economy; about how America will rebuild itself to new and undreamed-of prosperity by building an economy based on alternative, carbon-free or low-carbon energies. We have already seen how problematic some alternative energy sources are, but that's only half of the problem. The other half is the fact that all these green energy companies are going to issue stock, borrow money and commit themselves to endless growth because they will function in the same economic paradigm that governs everything else.

They're screwed before they even get out of the gate, especially for the brief interval where oil will stay below $100. In the Peak Oil movement we have called this "The Bumpy Plateau" for more than a decade. Any attempt at economic recovery will result in an immediate oil price spike in the face of depletion, which will kill the recovery and take another, deeper bite out of what was left when the recovery started.

It would be unwise to instantly forget what happened with the dot-com and housing bubbles. Both were illusions and well-orchestrated wealth transfers from the middle and lower classes to the wealthiest people in the country. The housing bubble was created and fanned white-hot by intentionally deregulating the mortgage industry, fraud and a host of crimes which sucked people into buying homes they could not afford and could never hope to pay for. A ton of money was created and it went to the people who ran the schemes: the largest banks, mortgage lenders and political campaign donors.

When that bubble collapsed, the taxpayers were asked to bail out first Bear Stearns and then Fannie Mae and Freddie Mac at total costs that will top $1 trillion dollars before counting the October 2008 bailout of $800 billion and all those that followed under many deliberately confusing names into the first quarter of 2009. As I write, the total "value" of various U.S. government bailouts has topped $10 trillion.

This doesn't count the U.S. banks that have failed and are going to fail before banks are inevitably nationalized. Those are the same banks where green energy companies will be forced to look for financing. Personally, I think that the sooner the big banks fail, the sooner people can get to devising local currencies, which is what they'll need to survive anyway. It is imperative to start that process while bridges are still standing and fresh water still runs. We need to start the transition to local currencies while there is still electricity and while fiber-optic cables are maintained and relatively new; while airlines fly and cell phones operate.

None of the above takes into account all the cash that home-buyers put into down payments initially. That money was lost too. That's the same thing as the money that gullible investors poured into the dot-com bubble. The ones at the bottom of the pyramid are always us, and it is always our money that disappears first. The current monetary paradigm offers no other option. The above does not address the equity (energy) that was lost in each collapse. These are real costs.

In the market crash of 2002 and 2003 (which I accurately predicted, saying it was only a precursor to today's events) hundreds of billions of dollars of shareholder equity were destroyed by the fraud of major corporations. Those dollars represented a lot more energy than what circulates today. The Federal Reserve has doubled its capitalization in less than a year, having left it alone for the previous nine decades. The equity was destroyed, but the wealth was transferred. And equity is where wealth resides in the dying economic paradigm.

There may be 40% less equity in the Dow Jones than there was in late 2007, but there is more equity that has been hidden and disguised by those who hold it. But even wealth transfers have a law of entropy. This is not a case where all those investments were converted 1:1 into some other form. The elites who thought they were immune are going down too, like dinosaurs who cannot grasp their impending extinction. Even the Oracle of Omaha, Warren Buffet, has discovered himself mortal.

As the networks blithely talked about shareholder equity that was lost at the beginning of the collapse, they almost never mentioned how many billions of dollars pension funds, other institutional investors and individuals put back in to the markets when they bought more shares at newly lowered prices. When bubbles burst, those on the bottom literally pay twice. The first time, when they buy stocks that later tank, and again when they purchase new shares, hoping to make up for the equity they lost when the previous bubble burst. Does this sound like an out-of-control gambling addiction to you? What happened was that the people at the top got "their" money out, at the top. They sold their shares before the bubble burst. That's why they call it "pump and dump."

An American president cannot let this happen with a "Green Economy" for three reasons. First, the Treasury is empty and the United States now has its largest budget deficit ever, with the national debt exceeding $11 trillion. It doesn't have many bailouts left, and these do absolutely nothing to solve the fundamental problem. They only impair the system's ability to respond to new challenges, like feeding you when the time comes. Second, the infrastructure costs to assist in some kind of stable transition and to maintain basic services as oil and gas fade away are going to be astronomical. Third, the Green Economy has got to produce and deliver useable solutions quickly. We cannot afford energy bridges to nowhere that make great profit for investors but provide little or no real-world benefit. If the Green Economy doesn't do this, then the nation will be left with a non-functioning energy infrastructure.

Beware of Greenwash hype.

A new level of oversight by the Securities and Exchange Commission (SEC), managed directly by the White House, is going to be essential. There will need to be the equivalent of a Good Housekeeping Seal of Approval for alternative energy companies which says that what they are selling will actually work. We know what to look for. The financial folks who will organize and fund the Green Economy will—as a matter of course—be of the same discipline, with the same priorities, trying to meet the same requirements as the folks who gave us Enron, WorldCom, Tyco, Bear Stearns, Fannie Mae, Freddie Mac, Lehman Brothers, Citigroup, AIG, and Washington Mutual. If the Green Economy is to be any real help, it must have, as its only mandate, the development and delivery of alternative energy supplies and infrastructure and getting it to the American people in an efficient and speedy manner.

This will require a fundamental change in the way money works, and it will be directly addressed in the proposed policies which follow.

"It May Not Be Profitable To Slow Decline"

In 2003 I reported on a conference of the international Association for the Study of Peak Oil (ASPO) in Paris. It was at that conference that a Dutch economist delivered an analysis of all the possible alternatives that were supposedly going to replace conventional oil. Here's what I reported.

> After looking at more of the various alternatives, [Martin] Van Mourik revealed an underlying truth that is certain to exacerbate the effects of Peak Oil, 'It may not be profitable to slow decline'.[5]

Simply put, more money can be made—more quickly—by accelerating decline, bankrupting the country, starving people, and selling off assets than by investing it in rebuilding under a new economic paradigm or by trying to soften the crash. The destruction of economies will also destroy demand for energy. Demand destruction has lowered prices which have made it unprofitable to invest in alternative energies. This is the path chosen and advocated by investment houses like Goldman Sachs, J.P. Morgan, and influential think tanks.

Think I'm mistaken about selling off assets? Do you recall that Anheuser Busch was recently sold to Belgian brewer InBev? As airlines fail, they are being liquidated and future mergers will generate more profit while reducing service. Two of America's big-three automakers have gone bankrupt. They face the wholesale liquidation of their assets for pennies on the dollar to better-positioned foreign car makers and liquidation firms that can ship entire industries overseas, piece by piece, making more money by stripping the carcass than they could by selling the company intact and saving its jobs.

So who's going to make all those electric cars everyone's talking about then?

Money is also time-sensitive. Quick profits and cash flow always trump longer term investment that may take years to pay back. Financial markets have no long-term vision in the infinite growth paradigm. If a Wall Street trader sees that more money can be made faster by putting something out of business, as opposed to waiting the five, ten or twenty years that alternative energy investments may need before showing positive returns, there is no doubt as to which choice will be made. This is especially true when there is little or no credit left and liquid cash is king.

The operating mantra for monetary decision makers throughout most recent history has been, "Make money on the way up. Make money on the way down." Energy is needed in both directions. But no one can "make" energy. The current economic paradigm will find that it uses less energy to make more money by driving things down than by building them up. It can't help itself, as in the parable of the turtle and the scorpion.

Feed-In Tariffs (FiTs)—A Humble But Powerful Beginning

A Feed-in Tariff or (FiT) is a national law that provides that those who install any kind of renewable electrical generating system (solar or wind) which produces more electricity than is consumed on site may sell the surplus back to the grid at above-market rates. These government-guaranteed rates can be as high as 40% above retail and locked in for a long-term period (usually twenty years or more). This takes all the risk out of the heavy up-front investment needed to install such systems. It reduces the amount of time needed to reach profitability and has caused an explosion in solar (photovoltaic or PV) and wind generation throughout Europe. It also gives an enormous boost to relocalization efforts. Nowhere has this been more dramatically demonstrated than in Germany which pioneered the concept.

There is no form of energy more important to industrialized civilization, to health or to life itself than electricity.

FiTs mandate that regional utilities buy all surplus electricity that is generated in this manner. Costs are not absorbed by utilities but spread among all ratepayers so the initial impact is barely noticed if millions of ratepayers are subsidizing the cost of these investments. In essence, the government guarantees that no one can lose money by investing in renewable electrical generation. After the development and investment costs are paid off, what remains is greatly expanded generating capacity, using less fossil fuel energy. Utilities still generate revenue from administering the process and distributing/selling the excess throughout the grid.

The resistance in the United States will come from giant utilities which can make more money, faster, by trading energy as a financial product from region to region when there are shortages. This serves no one except the financial markets. It is an inefficient use of energy. It does not produce new capacity and, in fact, continues the paradigm of reducing infrastructure investment and replacement to maximize short-term profits.

However, the overwhelming success of FiTs in Europe, including economic expansion, job creation and reduction in greenhouse-gas emissions looks certain to ultimately overpower any resistance in the United States . . . if only the United States government could force itself to admit that FiTs exist.

In Germany, as a result of FiTs initiated in 1999, newly installed PV generation capacity has soared from 16.5 Megawatt hours (MwH) in 1999 to 145 MwH in 2003. Ninety percent of that new capacity is tied into the grid, the rest being used in stand-alone systems. In addition, the German PV industry generated over 10,000 jobs in 2003 alone and added revenues of over 800 million Euros to the German economy.[6] Overall, the PV industry in Germany employs 40,000 people with an estimated total of 140,000 jobs in all renewable-energy disciplines.[7]

Feed-in tariffs have exploded German green electricity production, giving Germany (as of July, 2007) 12.6% of its electricity from renewables while in Britain, which has yet to implement FiTs, only 4.6% of its electricity is from renewables. That explosion in renewable solar energy was continuing not only in Germany but all over Europe until the start of the current depression.[8]

In the United States, California has just enacted a Feed-in Tariff, and other states are looking seriously at them. But to be most effective, a Feed-in Tariff should be a national law and uniform between all states. This would be the only way to prevent excess capacity being sold off from one region to another to maximize profits and having rate-payers in, say, California, subsidizing electricity that might be sold to Arizona, Nevada or someplace else for profit that would go only to the utility company. One of the biggest essentials in transitioning away from carbon-based fuels is to maintain economic and physical stability throughout the nation while it takes place.

In July of 2008 (as this book was being written), Representative

Jay Inslee (D-WA) introduced federal legislation calling for a national Feed-in Tariff. It will be interesting to see how it is opposed and by whom.

I will recommend other baby steps to break the back of the infinite-growth paradigm in the specific points of the program which follows. This transition cannot happen overnight but must happen as soon as possible.

There can be no tangible improvement in our chances for survival until the monetary paradigm is broken.

Post Script

In October 2006, Jenna Orkin wrote one of our last essays for *From The Wilderness*. It dealt with climate change, unrestrained growth and rising fuel costs in what was becoming the perfect storm we see unfolding today. In "Wagging the Dog: Economic Growth Leaves Water, People and Food Supplies in the Dust," Orkin looked at the disastrous impacts of climate change and rampant growth on food, water, health and the increasing number of displaced persons as it impacted less-developed nations in Africa and Asia. After noting that many thousands had committed suicide subsequent to being displaced by sprawl, she concluded that:

> The good news is, the economy thrives. Like Frankenstein's monster escaped from the lab, it cuts a swath through the global countryside, oblivious to the peasants cowering in its path. Let the world and its peoples collapse by the wayside, the economy will feed on the remains, knowing that its purpose in life is to grow, no matter what the cost nor to whom. [9]

For most Americans these developments were distant then. They are not so far away now.

CHAPTER THIRTEEN

Foreign Policy

> Events in the five-year period which began on September
> 11, 2001 will determine the course of human history for
> the next five hundred years or more.
> —*Crossing the Rubicon*, page 8.

The fundamental question posed by this chapter is whether it is
better to fight over oil or to find some other way to respond to
shortages which are certain to be devastating for all nations and
all peoples. The question that immediately follows is whether it
is even possible to find another way, especially given mankind's
history. Resource scarcities and competition between nations
over them have always led to wars in the past. In fact, that's what
wars have *always* been about.

Since 60% of the oil on the planet is in the Middle East, and
the U.S. imports 70% of its oil from there, Africa, Canada,
Mexico and Venezuela foreign policy cannot be overlooked.
Since nations like Japan, Korea, and India—which have no
oil—must compete for their oil in these same places, there is no
nation that is not affected. Energy is a global issue.

Two things have been pretty much enshrined in conventional
wisdom over the last seven years: U.S. foreign policy is driven by
energy concerns, and the invasion of Iraq had nothing to do with
weapons of mass destruction. It was about oil. George W. Bush,
the Federal Reserve, Britain, and 2008 Republican Presidential
nominee Senator John McCain have admitted that.

Here are the lead sentences from a *Bloomberg* story dated May
1, 2003.

London, May 1 (Bloomberg)—The U.S. and U.K. went to war against Iraq because of the Middle East country's oil reserves, an adviser to British Prime Minister Tony Blair said.

Sir Jonathan Porritt, head of the Sustainable Development Commission, which advises Blair's government on ecological issues, said the prospect of winning access to Iraqi oil was "a very large factor" in the allies' decision to attack Iraq in March.

"I don't think the war would have happened if Iraq didn't have the second-largest oil reserves in the world," Porritt said in a Sky News television interview . . .

The headline for the story read, "'U.S., U.K. Waged War on Iraq Because of Oil,' Blair Adviser Says."[1]

The *Washington Post* got the same thing from President Bush in November 2006.

During the run-up to the invasion of Iraq, President Bush and his aides sternly dismissed suggestions that the war was all about oil. "Nonsense," Defense Secretary Donald H. Rumsfeld declared. "This is not about that," said White House spokesman Ari Fleischer.

Now, more than 3½ years later, someone else is asserting that the war is about oil—President Bush.

As he barnstorms across the country campaigning for Republican candidates in Tuesday's elections, Bush has been citing oil as a reason to stay in Iraq. If the United States pulled its troops out prematurely and surrendered the country to insurgents, he warns audiences, it would effectively hand over Iraq's considerable petroleum reserves to terrorists . . . [2]

Tongue-in-cheek, I might have to ask if anyone who competes for oil supplies is automatically considered a terrorist. We will see how U.S. foreign, military and legal policy has implied that this might actually be the case.

Alan Greenspan, then Chairman of the Federal Reserve, has admitted that Iraq was about oil. Quoting his then-newly released memoir, Britain's *The Guardian* laid his words out for all to see.

> In his long-awaited memoir—out tomorrow in the U.S.— Greenspan, 81, who served as chairman of the U.S. Federal Reserve for almost two decades, writes: "I am saddened that it is politically inconvenient to acknowledge what every-one knows: the Iraq war is largely about oil."[3]

Here is Presidential candidate John McCain from May of 2008.

> My friends, I will have an energy policy which will elimi-nate our dependence on oil from the Middle East that will then prevent us from having ever to send our young men and women into conflict again in the Middle East.

I took notes on this while watching television. Ipso facto: we went to war because of oil.

So let's get this out of the way. Iraq had nothing to do with 9/11. Iraq had no weapons of mass destruction. Iraq was all about oil—and nothing but oil—from Day One. It's no secret. In fact, the entire world knows and understands this. So what, then, did it mean when President Bush directed his Cabinet and the military to begin drafting plans for the invasion of Iraq just six days after September 11?

A major 2003 story from *The Washington Post* contains some quotes that become explosive today, especially within the context of this book. The title of the story was "US Decision on Iraq Has Puzzling Past."

> On Sept. 17, 2001, six days after the attacks on the World Trade Center and the Pentagon, President Bush signed a

2½-page document marked "TOP SECRET" that outlined the plan for going to war in Afghanistan as part of a global campaign against terrorism.

Almost as a footnote, the document also directed the Pentagon to begin planning military options for an invasion of Iraq, senior administration officials said. . . .

"Saddam Must Go."

A small group of senior officials, especially in the Pentagon and the vice president's office, have long been concerned about Hussein, and urged his ouster in articles and open letters years before Bush became president.

Five years ago [which would be in 1998, or three years before 9/11], the Dec. 1 issue of the Weekly Standard, a conservative magazine, headlined its cover with a bold directive: "Saddam Must Go: A How-to Guide." Two of the articles were written by current administration officials, including the lead one, by Zalmay M. Khalilzad, now special White House envoy to the Iraqi opposition, and Paul D. Wolfowitz, now deputy defense secretary.

"We will have to confront him sooner or later—and sooner would be better," Khalilzad and Wolfowitz wrote. . . .

The Pentagon, while it was fighting the war in Afghanistan, began reviewing its plans for Iraq because of the secret presidential directive on Sept. 17. On Sept. 19 and 20, an advisory group known as the Defense Policy Board met at the Pentagon—with Rumsfeld in attendance—and animatedly discussed the importance of ousting Hussein. . . .

"I do believe certain people have grown theological about this," said another administration official who opposed focusing so intently on Iraq. "It's almost a religion—*that it will be the end of our society if we don't take action now* (emphasis added)."[4]

"The end of our society . . . "? Indeed.

Note that Khalilzad has served since 9/11 as U.S. Ambassador to Afghanistan, Iraq and currently the United Nations. Just prior to the 9/11 attacks Khalilzad was a special adviser and liaison between oil giant Unocal and the Taliban regime in Afghanistan. Those were the guys who provided a safe-country for al-Qaeda. His mission? Negotiate deals for pipelines from the Caspian basin through Afghanistan to get Caspian oil to market without going through Russia. Those negotiations collapsed just before the attacks.[5]

It was widely reported at the time that the Iraq focus was Bush's (Cheney's and Rumsfeld's) first concern immediately after the attacks and that it had been well deliberated long before the attacks. Countless documentaries and films, news stories and personal accounts revealed that Bush was almost "obsessed" with Iraq after 9/11.

As I described extensively in my book *Crossing the Rubicon*, no one in the military or intelligence communities was even faintly suggesting, or had any evidence whatsoever, that Saddam Hussein had anything to do with September 11. September 11, however, did provide the much-needed pretext for the invasion of Iraq as Bush and Cheney not-so-subtly implied that Saddam Hussein had been involved in it.

The World Trade Center and Pentagon—as the official story concluded—were attacked by terrorists operating from caves in Afghanistan, many of them carrying Saudi passports. "So then, it's obvious. Let's invade Iraq for its oil!" was the "evolution" of presidential thinking in the week after September 11.

In his 1980 State of the Union address President Jimmy Carter articulated what is now known as the Carter Doctrine when he said, "An attempt by any outside force to gain control of the Persian Gulf region will be regarded as an assault on the vital interests of the United States of America, and such an assault will be repelled by any means necessary, including military force." All Carter did was to state the obvious. In the case of post-9/11 Iraq there was no immediate threat to Iraqi oil from an outside country. The only problem with Iraqi oil was that it happened to be in Iraqi (Saddam Hussein's) control.

Is The Global War on Terror Really a War for Oil?

It's a fair question, isn't it? Assuming that the reader is American, ask yourself what the rest of the world might be thinking.

Since the attacks of September 11, the United States has tele-graphed to the world an aggressive and punitive posture. It has steadfastly refused to meet with opposing, or even truly inde-pendent, world leaders in an unprecedented show of diplomatic intransigence and inflexibility. In the National Security Strategy of the United States[6] released in September of 2002 the Bush administration set out a clear policy that was tough as nails. It set a tone in keeping with George W. Bush's "either you are with us or you are against us" message immediately after the attacks.

As I wrote in *Crossing the Rubicon:*

> Among its most elastic mandates for procedural omnipo-tence is George W. Bush's National Security Strategy of the United States.[11] That document describes two shocking de facto powers of the Empire: to launch without provo-cation, pre-emptive strikes anywhere it wishes against any nation that might someday be a threat; and to create artifi-cial terrorist activity where it wishes to deploy troops, with an avowed policy of lying to the world through unprece-dented manipulation of the corporate media with which it colludes. The Empire has thus defined the scope of conflict at the end of the age of oil: a no holds-barred, no-rules, and no-quarter race for global domination.[7]

As I described in *Rubicon* the United States then created the Proactive Preemptive Operating Group (or P2OG), which was empowered to preemptively attack suspect groups around the world. That was the largest expansion of U.S. covert operations since the Vietnam war.[8]

The United States also authorized "rendition"—a program of kidnapping suspected terrorists from neutral countries and flying them to "friendly" third countries where they could be "rendered"—or tortured—for information.

The United States passed the Patriot Act, which drastically reduced constitutional protections. It opened a detention facility in Guantanamo Cuba where "detainees" could be held indefinitely without charge and without the protections of the U.S. Constitution afforded to any criminal defendant within its borders, regardless of nationality. It created a $40 billion Department of Homeland Security; passed massive biological warfare legislation which allowed for the development of new weapons; and accelerated an ongoing expansion of U.S. military deployments in and around key energy-producing and transporting nations.

I specifically addressed the last point in *Rubicon*.

> "Also of primary importance would be any region that included a geostrategic oil transport route. As an example of the latter I would offer the Straits of Malacca and the South China Sea through which oil tankers supplying Japan, China, and Korea must pass. Others would include the former Soviet Republic of Georgia, Turkey and (for the future) Iran . . . "[9]

In the late summer of 2008, as this book was being completed, Russian retaliation over U.S. military, political, and economic expansion in Georgia has raised global military tensions to their highest point since the Cuban Missile Crisis of 1962. Georgia is a key transit country for the Baku-Tbilisi-Ceyhan (BTC) pipeline from the Caspian Sea to the Mediterranean Sea. Its roughly one million barrels of oil per day bypass all Russian territory; it was built for just that purpose. There are two other significant oil and gas pipelines in Georgia. As I documented in *Rubicon*, the United States began sending Special Forces troops into Georgia in 2002 and also became a sponsor for its entry into NATO, with the specific intent of encircling Russia with potential threats.

In August 2008, the Russian army suddenly invaded Georgia and occupied two "break-away" regions (S. Ossetia and Abkhazia). It is still—as of this writing—refusing to leave, and there is no real authority to make them. Georgia's admission process into NATO was thwarted, and President Bush, Vice President Cheney, and Secretary of State Condoleezza Rice have turned to rhetoric, followed by supporting action, which may lead to military confrontation and ultimately a nuclear option (as automatically dictated by Russian military doctrine).

The Obama administration has clearly and wisely recognized Russia's fait accompli and begun a rapprochement (reconciliation) with the only other country in the world capable of maintaining stability—in partnership with the United States.

I could expand this section a lot further, but I don't think it necessary. The point is clear. This is not a good environment for global cooperation on a matter of the gravest importance to all nations. As the "Big Dog" on the block, the United States has chosen the sheet music and the key, set the tempo and struck up the orchestra on a symphony which no one really wants to see reach its climax.

The United States of America must demonstrate that it will lead the way and make a needed sea change as Peak Oil worsens. If there is to be international cooperation it must begin with America. The United States must prove, both to the American people and to the international community, that it is not an aggressive partner intent on furthering "a war that will not end in our lifetimes" in the pursuit of hydrocarbon energy.

The Oil Depletion Protocol

There is a simple, straightforward proposal—easily read and understood—that could remove almost all risk of global confrontation over oil depletion. That proposal also would ensure that all nations carried an equal and transparent burden as Peak Oil's consequences continue to be both more and more obvious and severe. It is called

The Oil Depletion Protocol—it has been widely circulated for at least four years—and it was drafted by my friend and colleague, Dr. Colin Campbell, retired oil geologist regarded as perhaps the world's best-known spokesman on the subject of Peak Oil.

Rather than describe it. I will let it speak eloquently for itself.

THE OIL DEPLETION PROTOCOL
AS DRAFTED BY DR. COLIN J. CAMPBELL

WHEREAS the passage of history has recorded an increasing pace of change, such that the demand for energy has grown rapidly in parallel with the world population over the past two hundred years since the Industrial Revolution;

WHEREAS the energy supply required by the population has come mainly from coal and petroleum, such resources having been formed but rarely in the geological past and being inevitably subject to depletion;

WHEREAS oil provides ninety percent of transport fuel, is essential to trade, and plays a critical role in the agriculture needed to feed the expanding population;

WHEREAS oil is unevenly distributed on the Planet for well-understood geological reasons, with much being concentrated in five countries bordering the Persian Gulf;

WHEREAS all the major productive provinces of the World have been identified with the help of advanced technology and growing geological knowledge, it being now evident that discovery reached a peak in the 1960s, despite technological progress and a diligent search;

WHEREAS the past peak of discovery inevitably leads to a corresponding peak in production during the first decade of the 21st Century, assuming no radical decline in demand;

WHEREAS the onset of the decline of this critical resource affects all aspects of modern life, such having grave political and geopolitical implications;

WHEREAS it is expedient to plan an orderly transition to the new World environment of reduced energy supply, making early provisions to avoid the waste of energy, stimulate the entry of substitute energies, and extend the life of the remaining oil;

WHEREAS it is desirable to meet the challenges so arising in a co-operative and equitable manner, such to address related climate change concerns, economic and financial stability, and the threats of conflicts for access to critical resources.

NOW IT IS PROPOSED THAT

A convention of nations shall be called to consider the issue with a view to agreeing to an Accord with the following objectives:

- to avoid profiteering from shortage, such that oil prices may remain in reasonable relationship with production cost;
- to allow poor countries to afford their imports;
- to avoid destabilizing financial flows arising from excessive oil prices;
- to encourage consumers to avoid waste;
- to stimulate the development of alternative energies.

Such an Accord shall have the following outline provisions:

- The world and every nation shall aim to reduce oil consumption by at least the world depletion rate.
- No country shall produce oil at above its present depletion rate.
- No country shall import at above the world depletion rate.
- The depletion rate is defined as annual production as a percent of what is left (reserves plus yet-to-find).
- The preceding provisions refer to regular conventional oil—which category excludes heavy oils with

> cut-off of 17.5 API, deepwater oil with a cut-off of 500
> meters, polar oil, gas liquids from gas fields, tar sands,
> oil shale, oil from coal, biofuels such as ethanol, etc.

Detailed provisions shall cover the definition of the several categories of oil, exemptions and qualifications, and the scientific procedures for the estimation of Depletion Rate.

The signatory countries shall cooperate in providing information on their reserves, allowing full technical audit, such that the Depletion Rate may be accurately determined.

The signatory countries shall have the right to appeal their assessed Depletion Rate in the event of changed circumstances.[10]

Another of my friends and colleagues, Richard Heinberg, Ph.D.—author of many books on the subject of energy—has described the obvious benefits of adopting the protocol.

"Reducing production and imports in this fashion would yield a number of advantages, not all of them immediately obvious.

"First, it would conserve the resource. Petroleum Engineers are keenly aware that oilfields that are depleted too quickly can be damaged, resulting in a reduction in the total amount eventually recoverable. Voluntarily and systematically reducing the rate at which the world's oilfields are depleted would extend their lifetimes, so that future generations could have access to a resource of which there is a finite quantity and that is useful for a wide range of purposes—both as a machine lubricant and as a feedstock for the production of pharmaceuticals, chemicals, and plastics—other than simply as a fuel.

"Second, from the standpoint of the participating nations, the reductions would be gradual and foreseeable. Nations, municipalities, and businesses would be able to plan their economic futures with minimal concern for dramatic price variations for oil and oil products, since there would likely be more world spare petroleum production capacity than would be the case without a protocol, and thus a greater ability to adjust to short-term causes of shortages—including geopolitical conflict, accident, and natural disaster."[11]

Easier said than done.

Tribunals and Secretariats do not appear out of thin air any more than energy does. They require negotiations, agreements and funding. But the world has come together quickly before and it can do so again, if the will to do so is sufficient. What is lacking is the leadership, and on this issue all eyes should rightly be focused first upon the President of the United States. By endorsing and agreeing to abide by the principles of the protocol, the United States would instantly affirm the principles of fair play which it professes, surrender any claim to supremacy or special prerogative, and lend its prestige and power to creating a place for all other nations to stand. I have always hoped that these were things that the United States really stood for.

From that place, true international cooperation and collaboration would be possible on every vital aspect of the energy crisis, including population growth and food.

On its present course the United States and the world as a whole are heading for increasingly dangerous military confrontations. These wars are, in fact, inevitable unless something fundamental changes. Otherwise the final solution to Peak Oil might indeed be nuclear.

There is, however, one fundamental weakness in the Oil Depletion Protocol. It is based upon and assumes accurate and audited knowledge of depletion rates. It is impossible to calculate rates of depletion with having accurate and trustworthy reserve numbers.

CHAPTER FOURTEEN

Setting The Policy

Based just upon the hard data in this short book, the first conclusion a president or any elected official should reach is that oil production will decline too quickly to provide ample time for a switchover to alternative sources. This is especially true for transportation, where there are no real alternative sources at all. The choices we have been told about are partially or totally inadequate, untested, or exist only on paper. Non-transportation energy is serious enough by itself, although some solutions like wind and solar are immediately applicable there. We can clearly see that what several studies told us is true: We should have begun this conversion in earnest in the 1970s, just as President Jimmy Carter recommended.

Oil is the first and paramount problem; and from 2009 on, the planet and the nation are looking at increasingly larger gaps between supply and demand. They will be on the order of millions of barrels per day, and the gaps will widen every year from here on out.

OK, so now what do we do?

I repeat that I am not a Democrat or a Republican, and the fact that I am expected to chose only between these two sorry alternatives infuriates me. I feel the same way about being offered choices between Capitalism, Communism or Socialism. Who says that these are the only options? Capitalism, Socialism, and Communism are constructs of the 18th and 19th centuries, and they all file on one crucial point: They all assume infinite resources. This has nothing to do with any ideology and everything to do with the nuts and bolts of keeping things running; of keeping people alive.

We have seen that switching to alternatives and the Oil

Depletion Protocol both share one important concern and outcome: the conservation or stretching out of the oil that remains. No one, not even the most die-hard, "drill-everywhere" conservatives are suggesting that a return to $1 gasoline and the massive increase in consumption that would follow are good things.

A return to $1 gasoline would be tantamount to giving a heroin addict a massive dose while trying to clean him up. It would only reaffirm and more deeply entrench the addiction.

Those days are gone. Get over it.

In addition, in all "reform" movements there is always a huge gap between theory and implementation. Would-be reformers tend to throw out theories as if they could be seamlessly adopted from thought without looking at the political and economic obstacles that always emerge—like so many gremlins—along mankind's "rosy" path to progress.

Nowhere in American governments are the consequences of that disconnect more obvious than—to use a cliché—where "the rubber meets the road." That would be in the offices of the nation's mayors, city councils and county executives. These are the leaders and managers who bear the biggest brunt of not only unrealistic expectations, but also of national fiscal mismanagement and corrupted, short-sighted, or self-serving policies. There's an old saying that shit rolls downhill. My favorite saying, however, is that "In a ham and eggs breakfast the chicken is involved, but the pig is committed."

These locally elected executives are the largest providers of service in the nation. They are the first responders and they are already suffering because of poor federal planning. It is at the local level where the first real bites of Peak Oil are already producing howls of pain, and that pain is going to get worse in very short order.

Any policy, whether it concerns energy or not, is obligated to take on the management of its implementation as well. If federal policy needs input or support from states, then the federal government is the logical point of coordination. It must deal

with people—their wants, needs, biases and also their igno-
rance, petty rivalries and jealousies. It must see and recognize
bureaucratic obstacles and financial limitations. Ultimately it
must make choices that cannot "please all of the people all of the
time" as Abraham Lincoln said.

However, tough choices about which services to maintain and
which to cut back should and must be made locally. An entirely
new kind of relationship between federal and state and local
governments must emerge quickly if the Union is to endure.

One of the first thoughts I had was that if the federal government has a Strategic Petroleum Reserve (or SPRO, as it is called)
to release oil into the economy as a whole, what do states and
cities have as a cushion? It turns out that's a very difficult question to answer because apparently no one, not even the U.S.
Department of Energy, has taken the time to find out. I placed
two calls to the DoE, asking for this information, and they were
not returned.

There is, however, some anecdotal data after Hurricanes
Katrina and Rita that should sound a loud alarm.

Daniel Lerch of the Post Carbon Institute, in a late 2007
book entitled *Post Carbon Cities: Planning for Energy and Climate
Uncertainty* wrote the following after interviewing Kathleen
Leotta, Lead Transportation Planner with the planning and engineering firm Parsons Brinckerhoff:

> In studying the shutdown of oil pipelines in North Carolina
> following Hurricane Katrina in 2005, she found that many
> municipalities were left to fend for themselves when their
> oil stopped flowing:
>
>> "A huge amount of their motor fuels were cut
>> off; they didn't quite seem to realize how much of
>> their finished fuels came through the pipelines.
>> The state held the largest stockpiles of fuel, and
>> when all the municipalities came to them to ask if
>> they could give them some of their fuel, they said

they couldn't because they didn't have enough for
their own vehicles and fleets.

It's really the case that municipalities need to start
thinking about some of these things on their own."

"Natural disasters are unusual and extreme events,
but this story nevertheless has a valuable lesson for local
government leaders: Know your municipality's vulnerabili-
ties, because there isn't necessarily anyone else thinking
about them."[1]

Rethinking Strategic Petroleum Reserves

In concept and practice, the Strategic Petroleum Reserve of
the United States is a cushion originally intended to protect us
against a temporary supply disruption as occurred in the 1973–4
OPEC Embargo. It was not designed to deal with permanent
shortages. It is intended to stockpile enough crude oil so that,
in the event of a supply disruption, oil can be released back to
oil companies and refiners to prevent a follow-on disruption in
gasoline supply. The oil companies are required to replace what
they take when circumstances permit. The currently approved
size is 1 billion barrels, but the SPRO now contains less than 700
million barrels, which would be about a seven-week supply at
current consumption levels.

I was a rookie police officer living in Los Angeles when the
73 "Arab Embargo" hit. The rationing that followed, along
with economic impacts that lasted for a decade, were devastating. In the Los Angeles Police Department we were ordered to
park our cars and sit in them for hours at a time to stretch fuel
stocks. Crime soared. The bad guys knew we couldn't/wouldn't
respond to maybe 50% of the calls. Gasoline thefts were occurring everywhere, ironically because the thieves knew we didn't
have enough gas to arrest them. The rationing system then was
that civilian cars with even or odd license plate numbers could
purchase limited amounts of gasoline only on even or odd days.

That was great when every station had full-time attendants and enough staff to read the plates.

Many nations like China have, in recent years, started creating their own SPROs.

In May of 2004 the Dow Jones news service announced that China had agreed to build a national petroleum reserve.[2] By January of 2007 the U.S. Federal Reserve Bank reported that China's ultimate goal was to have 400 million barrels in storage. Here's what the Fed reported.

> China's rapidly rising dependence on foreign oil supplies has created anxiety among its leaders about the security of those imports, more than half of which come from the Middle East or West Africa. China's apparently very modest commercial storage capacity probably has contributed to this concern.
>
> In view of these facts, China has followed other nations in establishing strategic oil reserves. China's long-run goal is to store 90 days of net imports, about 400 million barrels at projected future import rates. This would bring them into compliance with the International Energy Agency's (IEA) recommendation for strategic reserves.[3]

The U.S. strategic reserve is by far the world's largest. According to Wikipedia, Japan had, as of 2003, 320 million barrels in reserve. Japan is smaller than the United States, but it has no domestic production at all. The European Union has a more stringent requirement for each member country, requiring 90 days of total use per country, not just imports.[4]

It seems that everyone knows that oil is running out and has been planning for it. Too bad no one let us know.

Under a May 2001 agreement, all 26 member countries of the International Energy Agency are required to have a strategic reserve equal to 90 days of imports. After attending international oil conferences in Paris, Berlin and Edinburgh (and sending a staff writer to one in Lisbon) I can tell you that IEA requirements are not strictly enforced. At current import rates of just over 15 Mbpd, that would be about 1.5 billion barrels for America. But remember that U.S. production is in steep decline and the illusory bonanza from new unrestricted drilling won't start trickling for nearly ten years. No one realistically expects that to come close to offsetting declines in domestic production over the same period.

In the current system all disruptions are assumed to be only temporary. That's not wise is it? Oil borrowed from the SPRO, after release by the President, must be replaced by oil companies once the *temporary* disruption ends and the situation returns to normal. There are some obvious problems with this approach.

First, given depletion rates and demand uncertainty, it is a weak bet to assume that things are going to return to "normal" or that crude oil can or will be replaced at a significant rate over and above consumption.

Second, crude oil is released into the general marketplace through oil companies, which refine it and then sell it to distributors. Whoever has the money gets the gasoline. None is set aside for public safety or state and local government use.

Third, what is being stockpiled is crude oil, not gasoline or diesel. In a major societal or economic breakdown, what will be needed is refined product, not crude.

In 2005 the U.S. Secretary of Energy was directed to fill the SPRO to its previously authorized one-billion-barrel capacity. These figures do not include what the DoE calls a national defense fuel reserve of an unspecified size. According to the DoE web site, that filling hasn't been done yet. In January 2007 President Bush "announced that the SPRO should be expanded to 1.5 billion barrels in order to increase the nation's energy security."[5] Five years after implementation the United States has not complied with the IEA mandate. Many other included nations aren't complying either.

This is one more clear indication that the Bush administration has been aware of Peak Oil for a long time. One of its first acts was to create the National Energy Policy Development Group (NEPDG) under Vice President Cheney. Its findings, after two Supreme Court battles, remain classified to this day. In retrospect it looks as though what Cheney et al were planning for *was* Peak Oil—precisely what I started saying in late 2001.

Location, Location, Location

The entirety of the U.S. Strategic Petroleum Reserve is located in secure and sturdy salt caverns along the Gulf Coast. But while the oil may be secure in those caverns, the coast itself, and its precious infrastructure (including refineries), is not. Both in 2005 and in 2008, after hurricanes Gustav and Ike, the oil infrastructure along the Gulf Coast, along with all pipelines emerging from it, was shut down. Without debating whether climate change is manmade or not, it is nonetheless very real. The hurricanes have taught us that, if nothing else. So have the melting of the polar ice caps, the droughts, the vanishing glaciers and the rising sea levels. So I ask, is the Gulf Coast the best place to put all of our petroleum eggs? It is clearly the one place in the country that is most threatened by climate change.

There are now four major issues:

1. The SPRO isn't big enough.
2. The SPRO contains only crude oil, not refined products.
3. The SPRO is not designed to protect state or local governments.
4. The SPRO is not geographically diversified and is currently in an environmentally threatened location.

The United States also maintains The Northeast Home Heating Oil Reserve which is two million barrels of emergency fuel for homes and business that has already been refined. A note on the Department of Energy web site indicates that this reserve is intended to protect against supply disruptions and also that "35,000 barrels were sold in 2007 for budgetary reasons. *The government will use the proceeds from the sale to repurchase a quantity of heating oil in the future.*"[6] (emphasis mine)

After the economic crash of 2008? After trillions in bailouts? And pay for it with what? Where did *that* money go? The oil they sold may have cost between $60 and $80 per barrel. How much will the government be able to buy back if oil is $100 or more? $150?

The federal SPRO is more important than ever. It must never be used for political purposes or as a slush fund. It has a different and essential kind of power in its perceived authenticity and trustworthiness. It is things like authenticity and trustworthiness that hold America together, and sometimes they seem as hard to find as energy.

Solution: Create A Second Strategic Reserve Of Refined Product For State And Local Use

After realizing all this, I questioned whether it might not be better to set up another SPRO of 750 million barrels, not for eventual public consumption, but for essential services, public safety, and to maintain government operations in the face of decline. In order to do that, it would have to be refined product. It would have to start now; and the filling of the first SPRO (or SPROI) would have to be stopped where it is. The IEA needs to come up with a better plan.

Take all of this new production that is expected after repeal of the offshore drilling bans; take everything that comes from the Arctic National Wildlife Refuge (ANWR); take oil from everywhere it can be bought and chop it into thirds. As a tax on domestic oil companies which have recorded record windfall profits in recent years, mandate that one out of every three barrels produced must be given to the United States government to fill a second Strategic Petroleum Reserve (SPROII) with refined gasoline and diesel. That reserve will belong only to participating state and local governments.

There is no time to waste. Since these new fields will not be productive for years, pass a law requiring an immediate tariff of 15% of all oil from large domestic corporate producers with revenues of over $50 million per year, and a 5% tariff on smaller independent producers. (My $50 million figure is arbitrary and would need to be researched further. But it seems a fair demarcation given heavy capital investment involved in production.)

Specify that participating state and local governments will have to pay the cost of refining, post-refinery transportation, fuel stabilization, storage and security costs on their own; but essentially give them the oil for free. This has several advantages. Refineries are scattered throughout many parts of the country and thus not all vulnerable to hurricane damage in the Gulf. The storage would be at the state and local levels so it would be geographically diverse and readily accessible. State and local governments could shop for the best price among refineries and also best allocate their oil between gasoline and diesel as needs dictated.

To decide who gets how much, start with the best census data on population, then fix an equal and fair apportionment based on that, so that every region and every citizen knows they are getting their fair share.

The benefits for state and local governments are obvious. They are already broke under existing fuel costs and a collapsing economy. This would save them a lot of money and provide managers with solid data for contingency planning. It would also hedge against price instabilities that are driving markets and purchasing agents crazy today. It would guarantee that if decline becomes critical, emergency and vital services would continue so that society might function while adjustments were made. Buses, school buses and trains would run. Police, fire and emergency medical services could operate. Water pressure would remain. Streets would be cleaned and the trash would be picked up.

Many large multi-national corporations maintain their own strategic reserves. Why not cities and counties? Only the federal government can prime that pump.

Establishment Of The Policy

In the chapters on foreign policy, money, and here, in our last discussion, I have raised monumental, and in some cases seemingly unsolvable issues and challenges. In each chapter I have proposed one small way to begin solving each great problem. We take one step and a new one will inevitably appear, either out of choice or out of necessity.

It is apparent that in order to deal with Peak Oil and energy shortages, we, the people of the United States, need to rethink our government again, and in a way that has never been asked of us. That is our right and our obligation as citizens. Throughout our history America has always been able to fall back on what was once viewed as a whole "New World" of untapped resources, where "life, liberty, and the pursuit of happiness" were seen as infinitely expandable commodities. As consumers we have been taught that the only path to happiness was to consume more. In many ways, our constitution and the evolution of the executive branch have been incorporated and organized around that fundamental belief: a "work ethic" which says that you can have as much as you want—all that you want or can imagine—if you work hard enough for it.

Like most Americans I share a deep and profound distrust of the executive branch, the Congress and the courts as they are functioning and failing in front of our eyes. I can also see that the United States government is operating like a bakery trying to make drill bits.

As emergencies of every sort accelerate, and as public confidence sinks, the inevitable possibility of civil unrest and revolt move quietly forward from the rearmost ranks of our worry. It happened during the Great Depression. It happened in the 1960s. It can happen again. I would agree with Plato that order is the first requisite for some kind of positive outcome. However, I would be (and have been) as rebellious as Tom Paine at the sight of an authoritarian, unconstitutional or totalitarian regime in this country.

Whether as a result of the collapse of industrial civilization, the total insolvency of the U.S. government, the economic crash that is just beginning, or climate change (or all of the above) the effects on the U.S. government are going to be profound. Services are going to be cut. Bureaucracies are going to remain that may not be needed any longer, while some that may be desperately needed will not exist. At some point it will become necessary to put everything on the table and reorganize the government entirely. The imperative first-place-to-begin is the Executive Branch which should be created anew from a blank sheet of paper. I believe it is broken beyond repair and we have paid a steep price for it.

The post-Depression regulatory apparatus did not prevent the economic collapse of 2008. It became the vehicle by which a serious infection was (I believe deliberately) transformed into terminal cancer. However the U.S. and world economies emerge from the crises of 2008, they will never look the same as they did before it. Not surprisingly, based upon information from Britain's Oil Depletion Analysis Centre (ODAC), a 2004 analysis clearly suggested an economic implosion in 2007–2008 when it became clear—again based upon simple arithmetic—that severe oil shortages were inevitable starting in 2007. Of course, my newsletter *From The Wilderness* told our readers that in a story we published in February, nine months before ODAC evaluation of hard numbers confirmed the ugly truth[7] and prompted a press release in November, which we immediately published.[8]

In June of 2006 I published *FTW*'s fourth and last economic alert which warned in the strongest possible terms that the economic crash was coming. At the time I gave specific advice to all who would listen. Here is what I recommended:

- Wherever possible, stop spending money with major corporations trading their stock on Wall Street. Dump your AOL email account and go with a smaller company. Stop feeding the tapeworm, because now you are only enabling it to eat you faster. What we have told you here is that there are no longer any "biological" restraints to keep it from eating you completely.

- Withdraw your 401(k) and other investments from large, U.S. dollar-denominated stocks. Whatever you lose doing this will be nothing compared to what's coming.
- Put at least some of your money into precious metals. (Pay no attention to gold's recent price drops. Gold is a long-term investment and still paying off handsomely.) The trend is still way up and any gold below $600 an ounce is a terrific buy. Silver is also a solid buy. I still see gold at $800 before year's end and possibly at $1,000 by the middle of next year.
- Restructure your portfolio into local investments, taking special care to open bank accounts with small locally-owned banks.
- Get rid of—by whatever means possible—credit cards from major banking institutions; especially those with variable rates . . .
- Check the web and your neighborhood and see what's being done with local and regional alternative currencies in your area. Regional and local currencies will save more lives than a thousand freeze-dried survival meals.
- Do whatever you can to support local agriculture and farming. Start a vegetable garden yourself.[9]

Those who followed my recommendations actually made money on September 29, 2008 when the Dow lost 777 points in one day. That began a meltdown which will continue to worsen. I do not see a bottom in the Dow until it reaches the 3000–4000 range, which is where it must end up after there are no more illusory paper options to suck more wealth out of investors—Pump and dump. I am still receiving thank you emails, letters and calls from those who understood the real problem and followed my advice. It will be well worth your time to go and read the entire alert to see what else I said about the causes of a collapse which I had seen coming since 2001.

What prompted our final warning was a move by President

Bush—through then-National Director of Intelligence John Negroponte—to allow undisclosed major banks and corporations to stop reporting accurate balance sheet figures to the SEC on the grounds of national security. This probably allowed Fannie and Freddie, Bear Stearns, Lehman AIG and others to become less transparent and to suck additional cash into the markets before the crash, which a blind person could have seen coming. I believe that result was intended. Pump and Dump.

I use the word probably only because the identities of the companies allowed to conceal their bookkeeping under this Bush directive are still a national secret under the Obama administration.

Eleven days after I published that alert the *FTW* offices in Ashland, Oregon, were burglarized and all seven of our computers were smashed. I could attribute that to the economic alert, but we were also, at the time, publishing a seven-part investigative series in the tragic friendly fire killing of the brave former pro-football star Patrick Tillman. Tillman was serving with an Army Ranger platoon in Afghanistan and had become a staunch critic of the war while still heroically performing on the battlefield. We had the direct support of the Tillman family and 2,000 pages of Army records given us by Mary Tillman. Our work eventually resulted in the disciplining of nine officers, three of them generals, and the resignation of Defense Secretary Donald Rumsfeld just as the cover-up, obstruction of justice, and falsification of records reached his door. All of the mainstream media work that followed was, in many cases, a virtual cut-and-paste of our series. Every congressional handout used by the family to obtain eventual hearings was a reprint of our original series. So, take your pick.

The economic crash I predicted is, of course, exactly what has happened. The housing market began its devastating implosion in the fall of 2007, just over a year after I thought it would. It really became inevitable as vulnerable homeowners started to crack under rising gasoline prices which impacted their ability to make their ballooning mortgage payments. As adjustable rate mortgages kicked in on top of rising oil prices in 2008, as oil soared to $147 a barrel, the wheels started coming off of the paradigm of infinite growth—which was, from the start, predicated on cheap energy and nothing else.

At some point the question will have to be asked as to whether it is better to keep patching an Executive Branch that has basically been in place since World War II or to just design and build a new one that more accurately reflects the end of one era and cares for the needs of the paradigm being born.

These are not short eras but changes in human history equivalent to what the Industrial Revolution did to the Middle Ages and Renaissance—in reverse. That's a topic that needs more input and study than just one person can give. It will need new Thomas Jeffersons, James and Dolly Madisons, Alexander Hamiltons, John and Abigail Adamses, Ben Franklins, and Thomas Paines. I believe they are out there and that they will emerge. I have met some of them. I am one of them.

We may see in our lifetimes a second Constitutional Convention or ConCon. Let us pray that we do not have to get there the way the folks who ran the first one did—on horseback.

I pray also that we do not see a second Civil War.

As short as this book may be—as flawed as some will find it—I know that it is far superior to any plan that has been articulated anywhere, or anything that Barack Obama has moved on since his inauguration. Barack Obama must be forced to reverse himself on what he calls "Clean Coal." The government and media continue to deceive us about alternative energies and what they can do. They either don't get it, or they refuse to admit that they *do* get it; and that is incredibly dangerous, especially given the historical events we are experiencing as I conclude this book.

ADDITIONAL INFORMATION

U.S. Department of Energy: http://www.doe.gov

http://www.postcarboncities.net.

CHAPTER FIFTEEN

An Emergency 25-Point Plan for Action

NOTE: For each of the following points it is assumed that each shall be carried out in a constitutional manner—that where necessary, laws will be passed and judgments rendered. However, I look back at Franklin Delano Roosevelt and Abraham Lincoln and pray that we can find that kind of American just one more time, in case we need him—or her.

POINT ONE

Create SPRO II

Create a second Strategic Petroleum Reserve of 750 million barrels of refined products for state and local governments.

Stop filling the current Strategic Petroleum Reserve at 750 million barrels. This will require going back to the International Energy Agency to renegotiate agreements. This is necessary anyway because most member countries have not complied with Strategic Petroleum requirements under the old treaty. The likelihood that they will be able to do so after the market meltdown of October 2008, and in the face of decline and collapse, is slim.

SPRO II will be created and reserved solely for the use of state and local governments.

SPRO II will not be centrally located but geographically diverse, with storage facilities selected, owned, operated, and maintained by participating state and local governments.

The federal government shall require and implement the following:

1. Pass a law requiring an immediate tariff of 15% of all oil produced from large domestic producers with revenues of

over $50 million per year, and a 5% tariff on smaller independent producers. Credit these oil tariffs at a uniform per-barrel price against corporate income taxes for producers as a form of price stabilization to ensure that temporary and short-lived gluts do not increase domestic demand by lowering prices below $90 per barrel. (These figures may need adjusting after detailed analysis.)

2. As new production, stimulated by the 2008 repeal of offshore drilling bans and the opening of the Arctic National Wildlife Reserve (ANWAR), comes online between (est.) 2016 and 2018, mandate that one in three barrels produced from these federal leases be diverted to SPRO II.

3. Producers shall be responsible for the cost of transport from wellhead to local refineries by participating governments as allocated by the Secretary of Energy.

4. State, city and county governments shall be responsible for the following costs:

- Refining crude oil into either diesel or gasoline in proportions dictated by jurisdictional needs.
- Transportation from refinery to local storage facilities.
- Selection, construction and/or purchase, and maintenance of local storage facilities in accordance with EPA rules and regulations.
- Fuel stabilization to prevent deterioration of fuel stocks in storage.
- Security. (It will be a federal offense to use SPRO II products for anything other than their intended purpose. Certain exemptions may be granted to local government executives to deal with unforeseen contingencies.)

5. State and local governments shall have the sole authority to allocate stored fuels or to share between governmental entities as needed.

6. It shall be a federal criminal offense to divert any fuels from SPRO II to commercial use unless there is a determination by each state and local government that such release is necessary to maintain vital functions and stability within their jurisdictions. Private firms contracted to state or local governments for the provision of essential services will be exempt. This will not preclude or preempt any state or local laws that may be enacted to protect and control supplies.

POINT TWO

Establish A New And Uniform Crude-Oil Reserve Accounting System For The United States

Through International Agreements With The International Energy Agency And The United Nations, Move For Uniform, Transparent And Reliable Reserve Estimation And Depletion Rates Globally

We must know how much energy there is and in what forms.

Direct the Secretaries of the Interior, Treasury and Energy to develop—based upon the best geological science available—a single, verifiable and transparent reserve accounting system for both oil and natural gas. The term Certified Recoverable Reserves (or CRR) shall be used to designate these reserve estimates.

All other designations, such as "probable reserves," "estimated reserves" and any other designations previously used for accounting purposes only shall be eliminated from geologic reports, SEC reporting requirements, and United States tax codes. There will only be one kind of reserve, and it will be as trustworthy as we can make it because our lives depend upon it.

Require that all energy-producing corporations and all downstream affiliated corporations use only Generally Accepted Accounting Practices for reporting purposes.

Make the results and the methodology used to compute reserve estimates publicly accessible as a basic right for all Americans.

Initiate immediate international action through the United Nations and the IEA, and on a multi-lateral basis with critical oil-producing nations, to adopt uniform and transparent reserve reporting requirements worldwide.

Recalculate depletion rates based upon the above and make them public.

Impose trade or economic sanctions on oil-producing nations refusing to comply.

POINT THREE

Enact The Oil Depletion Protocol

Use All Available Diplomatic And Economic Means To Encourage Global Ratification

WHEREAS it is desirable to meet the challenges so arising in a co-operative and equitable manner, such to address related climate change concerns, economic and financial stability, and the threats of conflicts for access to critical resources.

NOW IT IS PROPOSED THAT

A convention of nations shall be called to consider the issue with a view to agreeing an Accord with the following objectives:

- to avoid profiteering from shortage, such that oil prices may remain in reasonable relationship with production cost;
- to allow poor countries to afford their imports;
- to avoid destabilizing financial flows arising from excessive oil prices;
- to encourage consumers to avoid waste;
- to stimulate the development of alternative energies.

Such an Accord shall have the following outline provisions:

- The world and every nation shall aim to reduce oil consumption by at least the world depletion rate.

- No country shall produce oil at above its present depletion rate.
- No country shall import at above the world depletion rate.
- • The depletion rate is defined as annual production as a percent of what is left (reserves plus yet-to-find).
- The preceding provisions refer to regular conventional oil—which category excludes heavy oils with cut-off of 17.5 API, deepwater oil with a cut-off of 500 meters, polar oil, gas liquids from gas fields, tar sands, oil shale, oil from coal, biofuels such as ethanol, etc.

Detailed provisions shall cover the definition of the several categories of oil, exemptions and qualifications, and the scientific procedures for the estimation of Depletion Rate.

The signatory countries shall cooperate in providing information on their reserves, allowing full technical audit, such that the Depletion Rate may be accurately determined.

The signatory countries shall have the right to appeal their assessed Depletion Rate in the event of changed circumstance.

POINT FOUR

Immediately Declassify The May 2001 National Energy Policy Development Group (NEPDG) Records

Much of the indispensible work required in Point Two has already been done. The problem is that it is secret. Why?

In its first month in office, the Bush administration created a task force under Vice President Cheney to analyze America's energy situation in the context of both domestic and global conditions. With great controversy, arising after the Enron and other scandals of 2001–2003, the records, minutes and procedures of that task force were classified and withheld from the American people who paid for it.

A very small amount of records (7 pages) released after lawsuits, and two Supreme Court decisions indicated that the NEPDG had already undertaken many of the tasks identified as essential in Point Two.

In spite of a Supreme Court decision rendered after a private ex parte meeting between Vice President Cheney and Justice Antonin Scalia, who voted in the case, the withholding of this information from the American people is blatantly unconstitutional. American taxpayers paid for this report, and it will certainly contain a great many answers that are urgently needed.

If the United States is to demonstrate authenticity and good faith in meeting Peak Oil and energy shortages; if it is to regain high ground on the diplomatic scene, this action is essential.

If the people of the United State are to trust their government, this action is indispensible.

POINT FIVE

Impose An Immediate Moratorium On All Highway And Airport Expansion, Including Nafta Superhighways

Traffic and air travel are not going to expand as oil runs out. They are already decreasing. There is no point in destroying arable land, paving it with petroleum products and maintaining it for traffic that isn't going to be there.

Much more important will be the use of those funds and resources to maintain the roads and runways we already have.

POINT SIX

Completely Rebuild And Expand America's Rail System

On May 12, 2008, the Reuters news service reported the following:

In 2007, major freight railroads in the United States moved a ton of freight an average of 436 miles on each gallon of fuel. This represents a 3.1 percent improvement over 2006 and an astonishing 85.5 percent improvement since 1980.

"That's the equivalent of moving a ton of freight all the way from Baltimore to Boston on just a single gallon of diesel fuel," said Association of American Railroads President and CEO Edward R. Hamberger.

He noted that thanks to railroads' fuel efficiency gains, since 1980 freight railroads have reduced fuel consumption by 48 billion gallons and carbon dioxide emissions by 538 million tons.

Hamberger pointed out that railroads are three or more times more fuel efficient than trucks, adding: "In fact, if just 10 percent of the freight currently moving by truck went instead by rail, the nation could save one billion gallons of fuel per year."

Nothing will be more vital to the continuing function of the nation than this. In terms of cost and efficiency, nothing compares to rail transport. Since it is apparent that—whether we wish it or not—the United States will have to expand coal consumption as a new energy regime emerges. And it is imperative that there be enough rail transport to deliver it across vast distances.

In addition, the movement of other goods and people will be hobbled by fuel shortages and costs associated with less efficient means such as aircraft, trucks and automobiles. It will take decades to accomplish this task, and it should begin immediately.

AMTRAK should be fully funded and expanded to include not only major metropolitan areas but also smaller cities that are being cut out of commercial air passenger service in a shrinking economy with rising fuels costs.

The economy will depend upon this.

POINT SEVEN

Feed-In Tarrif

Immediately implement a national Feed-in Tariff (FiT) mandating that electric utilities pay 3% above market rates for all surplus electricity generated from renewable sources, especially including electricity generated by individual homes and businesses.

Nothing will more quickly ensure electrical energy security for all Americans.

POINT EIGHT

Speculation

All non-renewable energy sources in the United States shall be immediately classified as National Strategic Commodities. All derivative and index-based speculation in these commodities shall be immediately prohibited after a public date-certain with provisions made to clear out all pre-existing contracts by that time.

This will not include so-called "forward-hedged" purchases of oil, natural gas and coal by corporations or state and local governments. These are not speculative purchases but a means by which large consumers can adequately budget and avoid surprises, based upon anything other than actual supply and demand.

Price stability is essential to allow all governments and consumers to budget for energy needs.

POINT NINE

Enact A National Speed Limit Of
55mph And Strictly Enforce It

Immediately reduce the speed limit on the Interstate Highway System to 55 miles per hour and enforce it. This limit was imposed in 1974, and the country saved 167,000 barrels of oil per day in the following year. That lowered the nation's fuel consumption by 2%. It also reduced highway deaths.

The International Energy Agency's Workshop on Managing Oil Demand in Transport (2005) has estimated that the United States would save 4% of its total oil consumption by lowering the speed limit to 55mph. In 2008 American truckers actually supported a move to reduce the national speed limit to 65mph because of soaring diesel costs.

Provide that, on a state-by-state basis, speed limits may be returned to previous levels as soon as either 50% of all vehicles registered in a given state are either powered by non-petroleum sources or have CAFÉ mileage ratings at or above 45 City/55 Highway.

This will incentivize the use of rail transport and the manufacture and purchase of more fuel-efficient vehicles in a way that no other policy will.

If Americans are addicted to speed, then they can have it back when they find a way to go fast without oil.

POINT TEN

Eliminate All Federal Subsidies For Ethanol And Biofuel Production

As discussed in Chapters Eight and Nine, food supplies for all humans are in danger of declining as—or more rapidly than—available energy. The United States cannot and will never convert its entire corn crop (or any part of it) to ethanol production. The production of liquid fuels for transportation from either food crops or plant waste is extremely inefficient and harmful to the soil. Food is for eating. Plant "waste" is for replenishing the soil.

Eating is more important than driving. Unless drastic efforts are taken to heal our soil and make arable land available for projected population growth—or simply to offset productivity losses due to soil degradation—it shall be a cornerstone of U.S. policy that farmland be used to grow food (either for human or animal consumption) and nothing else.

Specific exemptions shall be granted, permitting the conversion of waste vegetable oils used for cooking and in specific cases where the Secretaries of Agriculture and Energy certify that plant waste cannot be used to enrich the soil within 50 miles of where it was produced.

POINT ELEVEN

Create Feed-In Tariffs For Local Food Production Point-Of-Origin Labelling

Direct the Secretaries of Agriculture and the Treasury to design a policy and propose legislation providing that any uncooked-food vendor shall be allowed to reduce his or her declared gross annual income by one percent for each one percent of food that is sold within 200 miles from its point of origin.

This will not include food products produced beyond that range and transported over distance solely for processing or packaging inside the range. In order to qualify, food sold must have its original source and be consumed within that radius.

Provide the same tax breaks for direct "farmer's market" sales of fresh, uncooked food at point of origin for local consumption or to fresh-food vendors and restaurants within the radius.

Subsidiary to this point, the Food and Drug Administration shall require all fresh-food vendors and manufacturers to clearly post the source country, state or county of all food items sold in their stores. This will require all food importers and manufacturers to provide this information and place it on labels, packaging and containers.

Mixed food-items shall be exempt from this requirement except in cases where the principal ingredient (more than 50% or largest ingredient by percentage of weight) is outside the United States. Manufacturers of canned goods and packaged fresh foods shall be required to clearly post country and/or state of origin on all labels before shipment to vendors.

Stimulate And Strengthen Local Food Production Through Federal And Local Governments And Make Vacant Urban Land Available For Cultivation

- Inaugurate a "Food, Not Lawns" program, incentivizing conversion of lawns to soil restoration and food production.
- After certifying that the soil is suitable and uncontaminated by toxic waste, make selected, federally owned vacant parcels available to local residents for farming without hindrance of any kind. Make this a "right of the commons" as fundamental as our Bill of Rights.
- Provide financial or other incentives to local governments which make vacant or abandoned lots available for cultivation.
- Prohibit monocropping on these parcels to protect the soil and reduce dependence on fossil fuels for fertilizer and pest control.
- Encourage the planting of native plant species for all landscaping within various climate regions as a means of soil restoration and water conservation.

POINT THIRTEEN

Agriculture—Soil Assessment

The Secretaries of Agriculture and Interior shall be directed with the utmost urgency to evaluate soil conditions around the country and to develop an emergency action plan for soil restoration and, wherever possible, the re-conversion of agriculture to organic means that are not fossil-fuel dependent. The Secretary shall evaluate overall food production rates and determine the most effective ways to transition without causing severe short- and medium-term disruptions in food production or supply.

After evaluation, the Secretaries shall prepare a detailed report for the President, constituting an inventory of all arable land in the nation, which breaks down soil conditions by soil type, health and productivity. The United States government shall publish this data immediately.

In addition, the Secretaries shall conduct thorough research into Permaculture, a food growing regime based upon steady-state, organic sustainability to support the American population as carbon-based energy supplies diminish. Organic farming functions without the addition of fossil fuel-based chemicals. Permaculture is an organic process intended to preserve soil fertility in perpetuity.

Monocropping shall be phased out nationally within ten years of the implementation of this policy to prevent further soil erosion and degradation.

For more information on permaculture, please visit the following three sites:

http://www.holmgren.com.au/
http://www.futurescenarios.org/
http://www.permatopia.com

POINT FOURTEEN

Create A Federal Clearing House To Track And Report On All Successful State And Local Intiatives With Respect To Relocalization And Energy Use

Direct the Secretaries of Energy, Interior, Agriculture and Defense to conduct an evaluation of all distinct regions in the country in order to identify:

1. Regions that are most vulnerable to energy shortages and any unique conditions which may exacerbate that vulnerability.
2. Regions that are most successful in their planning and adaptation with a view towards identifying local solutions and innovations that may be applicable to other regions.
3. Developing trends and problems that may need to be addressed at the federal level. Early warning of potential problems is critical to effective, efficient response.
4. Any changes to federal policy that might be needed either to correct ineffective policy or develop new policy in response to changing conditions.

No entity is better equipped than the federal government to perform this vital task. The quick, direct and essentially unfiltered dissemination of information to all citizens, who can act on it promptly, will be much more effective than the creation of a large, slow and inefficient federal bureaucracy. The federal government gathers the information and hands it directly to the localities to do with as they see fit, subject only to critical national interests as certified by votes of 75% or more in both Houses of Congress.

This information shall be published through the internet and by all other means deemed appropriate by the President.

POINT FIFTEEN

Draft And Pass A New Public Utility Holding Company Act

Re-enact a Public Utility Holding Company Act that will provide that all public utility companies must maintain sufficient energy reserves, infrastructure and resources to provide for ten-, fifty- and 100-year weather events, or other emergencies as determined by the Secretaries of Energy, Interior and Agriculture. This is essential because if something large breaks down later, rather than sooner, the chances that it will ever be repaired diminish greatly.

Mandate and enforce infrastructure repair and maintenance standards for all public utilities.

Prohibit energy trading between regional grids except on a non-profit, as-soon-as-possible payback between regions. This will guarantee that all energy providers are playing on one team rather than as competitors seeking to maximize profit at public expense, thus jeopardizing public safety.

POINT SIXTEEN

Rebuild The Grid And Energy Infrastructure, Including Oil And Natural Gas Pipelines

America's electrical generation and delivery system is falling apart due to profit taking and lack of investment. It must be rebuilt. The great Northeast Blackout of 2003 was caused by lack of investment in the grid, and the situation has grown worse since then.

Deregulation has further slowed maintenance, necessary repair and new construction. As part of an economic incentive package akin to what was done during the Great Depression, federal spending should be massively redirected away from military and other projects to protect this indispensible asset, put

Americans to work, and to ensure that continued economic functioning will be possible.

Oversight responsibility for this project should be given to the U.S. Army Corps of Engineers.

Create A Public Energy Oversight Board To Police And Monitor Advertising And Public Dissemination Of Information About Energy

It shall be a federal criminal offense to advertise, promote or sell any energy technology or regime that:

1. Is not immediately or imminently available for public use or application.
2. Is not net-energy positive.
3. Does not pass the tests outlined in this book.
4. Does not have an existing infrastructure capable of supporting or implementing it (e.g. "the hydrogen highway," where there are no service stations or pipelines capable of delivering an efficient fuel made from natural gas to vehicles that are commercially available in a significant quantity.)
5. Is deemed a severe environmental hazard by the Environmental Protection Agency and the Secretary of the Interior. The EPA shall be given such authority as part of this policy and the president shall personally supervise it.
6. Is a prohibitively heavy consumer of other strategic commodities such as fresh water, arable land, and forest or wilderness areas—all of which are in short supply.

Research and development of new technologies is essential. This will be especially true for the development of technologies that can increase the efficiency of energy use and conserve limited resources. But technology does not and will never replace energy itself. The laws of thermodynamics preclude this. Technology comes from energy, not the other way around.

Franklin Delano Roosevelt said that the only thing we have to fear is fear itself. We must also fear ill-conceived, false hopes and schemes which distract us into complacency instead of facing the crisis at hand. The same applies to any proposal which enriches confidence artists. The taking of investment and profit to pursue schemes which are not immediately applicable, which do not help, and which do not pass the tests outlined in this book is and should be a criminal act.

Any diversion of monetary and physical resources must be balanced between the need for immediate preparatory and responsive action and the benefits of possible long-term solutions. It does little good to expend our last dollars and hard assets to develop something that may have benefit in 20 years if there will be no serviceable roads or electricity in 10 because we put our resources in the wrong place.

On a case-by-case basis, new regimes with long "incubation" periods that offer real hope and have met the tests outlined in this book may be supported and exempted from these requirements, but only after a finding of "substantial benefit" by an impartial body. Rather than further defining that body here, I would rather leave it to a president and the Congress to deliberate and consult with the people, scientists and industry before creating it.

Whatever entity is created must be absolutely transparent and trustworthy; and it must be created in a way that will provide all Americans with confidence in its function.

The Federal Communications Commission, The Federal Trade Commission and other appropriate agencies shall have the authority to enforce the provisions of this section with a mandate that the protections of free speech are not curtailed. The protections of the First Amendment, however, do not permit fraud, confidence games, or utterances which clearly endanger public safety.

As a corollary, it shall be a federal offense to knowingly and willfully suppress any technology or discovery solely for personal gain, profit or influence.

POINT EIGHTEEN

Re-Draft The Tax Code Of The United States

Direct the Congress and the Secretaries of the Treasury and Energy to undertake a complete rewrite of federal tax codes.

The current U.S. tax code is archaic and thousands of pages long. It has been amended and modified over time to provide benefits encouraging (among other things) the growth and expansion of the old hydrocarbon-energy paradigm. This was appropriate for that era but it is not now. Although much of the oil depletion allowance has been pared back over the years, the tax code is still rife with provisions incentivizing coal, oil and gas production to the detriment of the development of alternative energy regimes.

The federal government must do everything possible to make it more profitable to produce alternative or energy-saving regimes. Today's need is to fully incentivize alternatives to carbon fuels without jeopardizing our ability to produce what oil, gas and coal we have left.

As a point of energy conservation, and to promote trust within the populace, it is recommended that an immediate flat tax (to be determined) be imposed for all income brackets. This will apply to corporations as well. The energy wasted in collecting, auditing, monitoring and enforcing the current tax structure—not to mention all that paper—is wasted.

Transparency is energy efficient.

Of all the commodities which are so essential in this time of crisis, trust between the people and the government and among the people themselves may be one of the most precious commodities of all. If the United States is to successfully deal with this crisis, then the American ability to come together in the face of danger must be nurtured at every step.

For accounting and regulatory purposes the SEC and the Treasury shall require all corporations to use only Generally Accepted Accounting Practices.

POINT NINETEEN

Nuclear Power

The enrichment of uranium for nuclear power and the construction of nuclear power plants is extremely energy intensive, expensive and time consuming. The waste produced from nuclear reactors is the most toxic substance known to man, and it remains deadly for hundreds of thousands of years.

It may be necessary to build more nuclear power plants to avoid societal and economic disruptions that threaten the nation's ability to function.

It shall be the policy of the United States government that all enriched uranium or other radioactive materials needed for such projects be obtained by dismantling currently existing nuclear weapons and using uranium that has already been enriched.

POINT TWENTY

Draft New Federal Building Codes For Home And Office Construction

Direct the Secretaries of Housing and Energy to draft new federal guidelines for the construction of all new homes and offices with a view towards maximizing energy efficiency and conservation.

Eliminate all codes prohibiting composting toilets and the recycling of rainwater and so-called "gray water" to encourage fresh-water conservation and local fertilizer production in such a way as is consistent with protecting public health.

Many countries around the world have been taking advantage of these options to great benefit for decades, without any problems.

POINT TWENTY-ONE

Education

The education of America's youth is now a matter of national survival. As local school boards are in some cases cutting back to four-day weeks and eliminating more expensive programs, the ability of the nation to meet future challenges is being diminished.

Education is essential to our survival, especially in the areas of science, mathematics, agriculture and social studies.

The Secretary of Education shall be directed to develop energy-curriculum standards in mathematics and the basic sciences at the middle school and high school levels. The specific intent will be to provide graduates with basic skill levels to meet infrastructure construction and maintenance requirements. This will also raise public awareness and begin a needed shift in cultural beliefs. Continued federal aid to education shall be contingent on compliance with these standards.

The federal government shall also fund—and the Secretaries of Education and Energy shall create and operate—a well-subsidized national vocational training program to produce the skilled workers necessary to implement wind, solar, organic farming, permaculture and other critical trades and skills as they are identified.

POINT TWENTY-TWO

Efficiency—Reduce Federal Government Energy Use By 15%

Except in cases where it would jeopardize public safety and emergency response, reduce energy consumption in all federal buildings and fleets by 15%.

If necessary, retrofit existing structures with energy-saving appliances and fixtures after careful analysis of net energy and cost.

POINT TWENTY-THREE

Drastically Reduce Overseas Military Deployment

Estimates vary, but open-source data from the internet indicates that the United States currently maintains between 700 and 1,000 military bases and installations in the United States and around the world at great expense and with huge energy costs.

In September of 2008, I attended the annual conference of the Association for the Study of Peak Oil—USA (ASPO-USA) which was held in Sacramento, California. Present at the conference were experts on energy issues from around the country. They came from the energy sector, industry, finance, and government. I have attended previous conferences and sent writers from *FTW* to others. It is clear that state and local governments are increasingly troubled by energy issues, and their attendance at this and related conferences is increasing rapidly.

One of the most prominent figures at the conference was Debbie Cook, Mayor of Huntington Beach, California, who has been a national leader on the subject and very effective at raising awareness among local leaders. During an interview Mayor Cook asked me, "Do you know that a U.S. Navy destroyer uses the same amount of oil in a week that the city of Huntington Beach uses in a year?" Huntington Beach's population as of 2008 is just over 200,000.

A reduction in overseas deployment of military personnel is inevitable. In all of recorded history, as empires faced energy shortages (long defined as shortages of slave labor), or as economic conditions deteriorated, military retrenchment was certain. Even before the economic crash in the fall of 2008 it was clear that energy shortages would require this anyway. This is not just a matter of cost, as oil prices dipped back into the $40 per barrel range as this book was completed. Depletion tells us that there will be as much as seven million barrels per day less oil available in 2009.

On November 11, 2009, *Wired* Magazine's Danger Room posted a story that opened with this paragraph. "Wanna know why the wars in Iraq and Afghanistan are so expensive? Here's one big reason: The U.S. military consumes 22 gallons of fuel per soldier, per day. And each gallon costs $45 or more to haul to the battlefield."[1]

The American Empire must shrink.

In order to protect this nation's security and to prepare for contingencies within the United States, it is an obvious step to undertake a complete review of U.S. military overseas deployments and operations and reduce them dramatically. Such reductions in force should be initiated only after careful review by the President, National Security Council and Secretary of Defense in conjunction with a completely new National Security Strategy addressing economic and energy issues together.

This is only an ill-informed estimate, but I believe that overseas deployments can and should be cut by as much as 30% to 50% without jeopardizing national security. Indeed, I believe that if these cuts are not made, the ultimate damage to national security could be much greater than if they are not.

To continue to expend that much money, and use that much irreplaceable non-renewable energy will jeopardize the nation's ability to function domestically. And it is the domestic economic health of the nation that makes possible military deployments in the first place.

Soldiers, sailors, Marines and airmen will want and need to be home to assist distressed families, and the nation may need them home to assist all of us in the coming years.

Especially since the attacks of September 11, the United States has jumped into the role of global policeman, projecting its military might virtually everywhere, to some degree. This is just no longer possible and, I would argue, no longer necessary.

As I have said so often over the years, in a post-petroleum world, geography will be the governing truth for international economic, military and diplomatic activity. We see this trend already. Latin America, Europe and Asia are slowly consolidating and looking for regional alliances. America is not exempt from this physical reality, this law of geography.

Let us decide now not to go the way of Ancient Rome, which had legions scattered over the known world as Rome itself was sacked. The chances of America being invaded militarily are slim to none. There is no need. We are being economically looted and ransacked from within, by our own monetary system. What will be left behind, however, as human civilization descends the curve of Peak Oil, suggests that we will need our sons and daughters home.

POINT TWENTY-FOUR

Decriminalize The Hemp Plant And Encourage Widespread Domestic Production

The hemp plant, also known as marijuana, is one of the most useful plants ever created by nature. I still laugh at the logic that says that mankind can outlaw a plant created by God and label it "bad." In fact, until beverage distillers and pharmaceutical companies trading their stock on Wall Street pushed to outlaw it, the U.S. government not only encouraged its growth but even subsidized it. During World War II "Hemp for Victory" posters were ubiquitous because there were so many great uses for something that could be grown in the back yard of almost every region of the country.

According to Jack Herer, the legendary advocate for hemp and marijuana legalization, listed in his 1998 book *The Emperor Wears No Clothes* many products of the hemp plant:

- Rope
- Textiles and fabrics
- Fiber and pulp paper (both the Declaration of Independence and the U.S. Constitution were written on hemp paper)
- Paints and varnishes
- Art canvas
- Lighting Oil
- Many medicines that do not depend upon pharmaceutical companies (Queen Victoria used it regularly for her menstrual cramps.)
- Food oils and protein (the seeds are a holistic health food full of protein and nutrients)
- Building materials

Nothing can replace oil or natural gas. But liberating the hemp plant will enable a limited amount of biodiesel production as fuel shortages grow worse.

Hemp needs no chemicals and has few weed or insect enemies—"except for the U.S. government and the DEA," according to Herer.

As a former LAPD officer who specialized in narcotics during the 1970s, I have come full circle on this point. I reject and refuse to participate in any hypocrisy on the subject. In the 1970s and 80s I was a mainstream guy—booze only. From 1983 through early 2004 I abstained from the use of any mind-altering substance of any kind. I didn't have even one beer.

I make no apologies or excuses about smoking marijuana today. I live in California yet I have no medical prescription that would allow me to smoke marijuana "legally." I will not lie about anything. For me that is a form of suicide. I finished writing *Crossing the Rubicon* while smoking marijuana two or three times a week and I wrote this book under the same conditions. I make no apologies to anyone and feel absolutely no shame.

Why the hell should I?

If Michael Phelps won eight Olympic gold medals smoking

it and if I can produce this work while using it, what in the name of God do I have to feel sorry about? I have not had so much as a moving violation on my record since 1973. I refuse to sell out to please anyone, and nothing intimidates me anymore. Although I do not use opiates in any form (unless having surgery or for a serious injury as prescribed), I feel exactly the same way about the opium poppy and the coca leaf (which I also do not use). It is not the plants which are harmful, but what mankind does to them for profit that is harmful. Cocaine hydrochloride and heroin (diacetyl morphine) are products of man's tampering with nature; no fault lies inherently in the plants themselves. Opium has been smoked for thousands of years because it relieves pain and eases tension. There's no good data anywhere that says smoke opium once, twice or even regularly and you become a murderous, rampaging criminal or addicted to anything else.

Arguing that smoking marijuana or opium leads to addiction and horrendous crimes is sheer scientific stupidity. I was a narc. The truth is (and always has been) that a certain percentage of the human population is genetically predisposed to addiction, whether to alcohol, cocaine, opium, heroin, tobacco or whatever. Addiction is a social, spiritual and medical issue and should be treated as such. This doesn't excuse crimes committed under the influence. The law has no problem locking people up for driving under the influence while allowing liquor stores every few blocks.

I have always been a lover of great, common-sense quotes. Here's one of my all-time favorites from retired Chief Joseph Macnamara of the San Jose Police Department. I heard it at a 1999 symposium on CIA involvement in the drug trade at USC.

"In 1972 when Richard Nixon declared 'War on Drugs,' the annual federal budget allocation was around $50 million. Today the annual budget exceeds $20 billion and yet there are more drugs on the street, they are of better quality and they are less expensive than they were in 1972."

What a stupid, ridiculous waste of energy and money . . . for Wall Street's benefit. If anyone needs to be released from prison, it should be all non-violent drug offenders first. But that's too simple. There are corporations housing inmates that derive their stock value from how many humans occupy their beds.

Liberate the hemp plant and saves lives. It's as simple as that.

(For the record, I took an honorary toke just to write this point.) Arrest me.

Initiate A Rational, Open And Ethical Domestic And Global Dialogue On Population Growth And Reduction

This may be the most difficult challenge ever faced by the human race. It will be by far the most important. It will also be perhaps the most difficult subject ever broached by an American president; but it must be.

The science and mathematics are unequivocally and universally clear. Given the economic collapse that is in its early stages as this book is finished, it is certain that starvation and global GNP reduction will have sharp impacts on population growth— much sooner than even I had anticipated. But by how much? Over what period of time? Will populations actually shrink, or will the rate of growth simply slow down?

At question here is not just the planet's ability to sustain new growth, which is obviously a thing of the past, but its ability to support those who are already here. We must power down.

There were only about two billion of us here before oil. There are almost seven billion of us today. Failing to address this single, overriding issue may result in the extinction of the entire species because, if we do not address this as a whole, it will be addressed for us by chaos, war, famine, disease, societal breakdown, collapse and very possibly nuclear war. This challenge may be addressed by those with vast money and resources in secret. It may be addressed by genocide, biological warfare or some other means.

These debates and decisions belong to all of us. I do not have a plan, but I am certain that there are people who do. It is time for the world to hear the thinking of the Peak Oil/Sustainability movement. I believe it may finally be ready to listen.

I have said for years that we are faced with a choice that can no longer be postponed or evaded.

Evolve or perish.

Adapt or die.

That is the universe in which all species live. Those are the rules that govern all life. We are not that special. We are not exempt. Our evolution must be one of consciousness.

Epilogue—Entropy

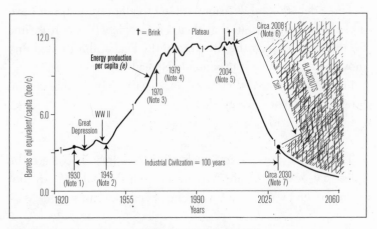

GRAPH COURTESY OF RICHARD DUNCAN, PH.D. PUBLISHED IN 2005. LOOK CLOSELY AT THE DATES.

Those who fondly believe that our economic growth will return any day now might contemplate what the Director of the International Energy Agency said earlier this week. His message is a simple one: investment in new oil production has fallen so low that whenever an economic recovery occurs, the oil to support the recovery is unlikely to be there. Supplies will be inadequate and prices will rise much as they did last summer, choking off the recovery.

Tom Whipple, "The Peak Oil Crisis: Parsing the Numbers,"
The Falls Church News Press, February 19, 2009

This will be the death knell of human industrialized civilization. I foresee it happening anywhere from the late summer of 2009 to the fall of 2010 . . . not much later. Since I have established an approximate 80% accuracy in my many predictions over the last decade, those in denial might consider getting angry and taking action now.

A November 13, 2009, story in Britain's *The Independent* confirms the predictions I started making in mid-2008. It suggests that we may not have even that much time to prepare. Headlined

"Surge in the price of oil is a fresh threat to recovery" the article's lead sentence said, "Another spike in oil prices could 'derail' the world's still-fragile economic recovery, the International Energy Agency warned yesterday."[1] What the story made clear was that even at $80 a barrel the price of oil was strangling any feeble moves towards recovery. Readers who have paid close attention thus far in this book will quickly grasp then that there is no recovery at all. There is only the paper illusion of recovery caused by an explosion of fiat currency. And with supplies declining faster than expected, coupled with the universality of fraudulent bookkeeping and criminally-misleading mainstream news reports on the subject, we may actually never really see a spike in demand before the collapse becomes irreversible and obvious to the whole world.

There has been only one true prophet who foresaw this moment clearly. All the rest follow on his trail. We have advanced and widened it some over the years. Physicist and geologist M. King Hubbert began warning of these days in the late1940s. In 1974 he testified about it at a House of Representatives hearing on National Energy Conservation Policy. Judging by events since, I can only conclude that Hubbert's words were taken very seriously by those who, over the last hundred years, have improved and enhanced their ability to manipulate public opinion and behavior for selfish or cowardly reasons. "Hell", as someone else wrote, "we all drank the damn Kool-aid!" The time when all of these lies, manipulations and self-deceptions are laid bare in the naked light of a stark and forbidding reality.

So what comes next?

Since I wrote the first words of this book, I have prayed for one thing and one thing only: that it might be possible to impart to as many individual human beings as possible the gut-level awareness of the magnitude of the crisis we face, and to enable those who do understand to prepare to face it, free of denial and with open eyes.

Colin Campbell, Ph.D. is perhaps the second greatest prophet who warned of this moment. He said to me once, "The species homo sapiens might not become extinct. But the subspecies of 'Petroleum Man' most certainly will." If the reader goes back to the human population graph at the end of Chapter One, by now it should be clear that Petroleum Man represents all of the population now in existence above the point in time when oil was first used in industrial civilization.

When King Hubbert testified before Congress in 1974, the human population was approximately four billion, with an annual compounding growth rate of about two percent. Today, as I write these words, human population exceeds 6.8 billion people. By the time this book has been edited, printed and released it will likely have topped 7 billion.

Let's go back and look at some of the things Hubbert told Congress 35 years ago.

> I was in New York in the 30s. I had a box seat at the depression. I can assure you it was a very educational experience. We shut down the country because of monetary reasons. We had manpower and abundant raw materials. Yet we shut the country down. We are doing the same kind of thing now but with a different material outlook. We are not in the position we were in 1929–30 with regard to the future. Then the physical system was ready to roll. This time it is not. We are in a crisis in the evolution of human society. It's unique to both human and geological history. It has never happened before and it can't possibly happen again. You can only use oil once. Soon all the oil is going to be burned and all the metals mined and scattered. That is obviously a scenario of catastrophe but we have the technology. All we have to do is completely overhaul our culture and find an alternative to money. We are not starting from zero. We have an enormous amount of existing technical knowledge. It's just a matter of putting it all together. We still have great flexibility but our maneuverability will diminish with time.

A *non-catastrophic solution is impossible unless society is made stable. This means abandoning two axioms of our own culture, the (current) work ethic and the idea that growth is the normal state of life.* Our window of opportunity is slowly closing, at the same time, it probably requires a spiral of adversity. In other words things have to get a lot worse before they get better . . . " [emphasis mine].

According to Hubbert, our window of opportunity was closing just a few years before President Jimmy Carter admonished us to wean ourselves from our gluttonous appetites. Somebody was listening, both to the message from Hubbert and the one from Carter.

Obviously, as documented throughout this book, our window of opportunity has closed, and a catastrophic die-off of human population is just beginning. One sign: There are more than 20 million unemployed in China. Many of them are displaced and far from home. More than 100,000 instances of civil unrest or rioting have occurred there since September 2008. At the same time a catastrophic drought is ravaging China's summer wheat crop, which has been infected by a new form of fungus. And yet the United States is expecting China to finance our endlessly expanding, multi-trillion-dollar bailout fantasies. Throughout 2009 China was still reporting near 8% growth (thereby increasing demand for needless consumer goods, while at the same time buying up as many of the world's commodities as possible to stockpile them.

To put it simply, recovery is what will kill us.

A second sign: My home state of California has just declared a devastating drought, and water to the state's farms is going to be cut back. Farms will shut down and harvests will shrink. And California is the largest agricultural producer in the United States. Worse yet, the collapse of the monetary system is collapsing food production all over the world for no reason other than to serve itself. The way money works is actually reducing global food supply because farmers around the world can't get the credit lines they have relied on for a century or more.

Yet the mainstream media will not connect the dots for you, or even place them in close enough proximity so that you can do it yourself. After all, they're just trying to protect the share prices of their corporate owners.

Of course, we could just have wars and riots. We could starve. We could die of thirst . . . or heat . . . or floods . . . or toxic waste. Some, I believe, have tried to make the decision as to who will live and who will starve for us. But the judgment of the dinosaurs who have ruled this epoch has been faulty all along, especially lately. Many elites of the financial class are going down themselves. Billionaires are an endangered species.

Life will continue, regardless.

In his written submission to Congress along with his oral testimony, Hubbert also wrote:

> Two terms applicable to an evolving system are of fundamental importance. These are steady state and transient state . . .
>
> . . . For more than a century it has been known in biology that if any biological species from microbes to elephants is given a favorable environment, its population will begin to increase at an exponential rate...
>
> Consequently when we compute a maximum average growth rate between the finite levels of population at a time interval of a million years, we arrive at the same conclusion, namely that the normal state, the state that persists most of the time, is one of an approximate steady state. *The abnormal state of an ecological system is a rapidly changing transient or disturbed state . . .*
>
> *Without further elaboration, it is demonstrable that the exponential phase of the industrial growth which has dominated human activity during the last couple of centuries is drawing to a close. It is physically and biologically impossible for any material*

and energy component to follow exponential growth . . . Yet,
during the last two centuries of unbroken industrial growth we
have evolved what amounts to an exponential growth culture.
Our institutions, our legal system, our financial system, and
our most cherished folkways and beliefs are all based upon the
premise of continuous growth. Since physical and biological restraints make it impossible to continue such rates of
growth indefinitely, it is inevitable that with the slowing
down in the rates of physical growth, cultural adjustments
must be made [emphasis mine].

In 1981, in a paper entitled "Two Intellectual Systems: Matter-
energy and the Monetary Culture" Hubbert wrote words *that* may
be the most profound observation in all of human history.[2]

The world's present industrial civilization is handicapped by
the co-existence of two universal, overlapping, and incom-
patible intellectual systems: the accumulated knowledge of
the last four centuries of the properties and interrelation-
ships of matter and energy; and the associated monetary
culture which has evolved from folkways of prehistoric
origin.

Having studied spirituality and religion for more than twenty-
five years, Hubbert's words answered for me perhaps the deepest
question of all. How did we get so screwed up? It seems that this
inevitable day of reckoning was chosen by the human race the
moment it left the Garden of Eden. Hubbert's use of the phrase
"folkways of prehistoric origin" clearly suggests this. It was on
leaving the Garden, with knowledge of "good and evil," that
we became trapped by pairs of opposites which have ever since
confused our thinking; rich/poor, good/bad, young/old, healthy/
sick. Money was the first hedge against fear and the father of all
subsequent efforts to avoid fear that have governed human civilization ever since . . . with compounding interest.

It was another of the many human prophets who have walked this earth who put the entire issue now confronting us as a species into just ten words.

"The love of money is the root of all evil."—Timothy 6:10

I find that same truth resident in the two spiritual paths which most heavily influence my life: Taoism and Buddhism. In fact, it didn't take too much digging to find this truth in every religion I have ever studied, including Christianity, Judaism, and Islam. I believe that this is a universal truth.

Until we change the way money works, we change nothing.

There are two issues about which I have never presumed to form or offer a solution. The first is what form money should take. The second is how to reduce human population.

As to the first, I can quote Hubbert.

> On this basis our distribution then becomes foolproof and incredibly simple. We keep our records of the physical costs of production in terms of the amount of extraneous energy degraded. We set industrial production arbitrarily at a rate equal to the saturation of the physical capacity of our public to consume. We distribute purchasing power in the form of energy certificates to the public, the amount issued to each being equivalent to his pro rata share of the energy-cost of the consumer goods and services to be purchased during the balanced-load period for which the certificates are issued. These certificates bear the identification of the person to whom issued and are non-negotiable. They resemble a bank check in that they bear no face denomination, this being entered at the time of spending. They are surrendered upon the purchase of goods or services at any center of distribution and are permanently cancelled, becoming entries in a uniform accounting system. Being non-negotiable they cannot be lost, stolen,

gambled or given away because they are invalid in the
hands of any person other than the one to whom issued.
They can only be spent. Contrary to the Price System
rules, the purchasing power of an individual is no longer
based upon the fallacious premise that a man is being paid
in proportion to the so-called "value" of his work (since it
is a physical fact that what he receives is greatly in excess
of his physical effort) but upon the pro rata division of the
net energy degraded in the production of consumer goods
and services. In this manner the income of an individual is
in nowise dependent upon the nature of his work, and we
are then free to reduce the working hours of our population to as low a level as technological advancement will
allow, without in any manner jeopardizing the national or
individual income, and without the slightest unemployment problem or poverty. [3]

According to Robert L. Hickerson, writing on a web site dedicated to M. King Hubbert's work, *"Hubbert goes on to state that
following a transition the work required of each individual, need be
no longer than about 4 hours a day, 164 days per year, from the ages
of 25 to 45. Income will continue until death. 'Insecurity of old age
is abolished and both saving and insurance become unnecessary and
impossible.'"* [4]

All of these calculations were made in the 1960s when infrastructure was new and everything was working. This is why it
is imperative that the transition to a completely new monetary
system begins while we still have functioning infrastructure. No
one really knows what the numbers will work out to be, but on
its face this concept is bulletproof.

Money and energy would now be free but permanently related,
forming not a pair of Siamese twins where the growth of one
kills the other; but a pair of separate, healthy and balanced twins
which could live together in both harmony and sustainability.

I strongly suspect that getting there is going to be quite a thrill ride with any number of possible outcomes. There has already been some violence, and there will be indescribable suffering and death to come. There will also be epochal revolutions in art, music and literature. There will be a complete testing and re-examination of all organized religion. From the human standpoint God will be on the table. In the meantime God (actual) will just be laughing as we painfully let go of a childish way of life . . . or not.

The key challenges will be not only how much survives . . . but what survives. I once asked my good friend Richard Heinberg, a world-renowned Peak Oil author and lecturer, how long the world could operate under the old paradigm, burning and consuming the way it did (even after Peak) if the population were reduced to around two billion. He shuddered as he thought about it. Then he kind of grimaced. "Oh, maybe another three or four hundred years."

Does anybody really want that? The human race is fighting not only for its soul, but very possibly its epitaph. A dying paradigm founded upon fear is fighting to recreate itself in another form by killing between three and four billion of us. The dinosaur part of that death-rattled struggle decided it was necessary to try to kill an emerging sustainable paradigm predicated upon balance and love for all life, and the planet we all call home. And yet I seem to recall another genius somewhere saying something like, "By the time anyone even recognizes the new paradigm, the old one's already dead." Let's hope he was correct.

I am not betting on the dinosaurs in this extinction. And this book is all about mitigating the die-off; for the children of the dinosaurs too. They will evolve even if their parents did not.

As to human population reduction, I have never once even dared to think of a solution. That is simply and solely because if ever there was a need for our species to actually talk honestly to itself it is now, about this subject. And it will be that dialogue— that soul searching—which will determine what survives after the collapse.

It's time to start that dialogue in earnest.

I finished writing this book in early September of 2008. It was perhaps one of the most difficult challenges of my life to perceive the urgency of the crisis and be compelled to watch and wait as the collapse of industrial civilization became not a future but a current event. Throughout the course of my writings over many years I predicted every aspect of the current economic crisis with an accuracy that only frightens me. The corporate media that said that no one could have foreseen this and politicians who echo that lie are rapidly losing all credibility. I may have been the most accurate but I was most certainly not alone.

There have been so many great warriors on the field of Peak Oil and Sustainability and I am honored to be in their company. I have not kept score but I have left a record of more than two million words proving this in four places. On the *From The Wilderness* web site; in my first book *Crossing the Rubicon: The Decline of the American Empire at the End of the Age of Oil*; on my blog at http://www.mikeruppert.blogspot.com; and in this book. There is no need to rewrite that. The only need is for people to read it.

All of us who tried so hard to awaken human consciousness before the collapse began can only share our grief and the frustration of going unheard and ignored . . . with each other. We have actually reached more than we may ever know. Over the years I have received an estimated 2000 letters and emails from families who understood this message and adapted before the crisis hit. It is never, ever too late for a human to adapt.

As I was awaiting the glacially slow and tortured process of seeing this book into print at a time when mankind's greatest challenge is taking its first painful bites, I looked for one simple metaphor to sum up our self-imposed moment of truth.

I finally found it where I usually find things after looking high and low for a long time . . . right in front of me.

Across the street from a dog park where I go every day with my dog Rags is a parking lot, largely hidden from public view, full of brand new cars—cars that will never be sold. As I drive around various parts of Los Angeles, I have found several of these out-of-the-way ad hoc storage lots, hastily rented since the start of collapse. They are all filled to capacity because so many dealers are going down and inventory has collided like an accordion. They are being used to quietly hold perhaps tens of thousands of brand new cars from Asia and Europe, each with seven gallons of oil in every tire, and many thousands more used in their manufacture; in the paints, plastics and resins. I think of the oil used to transport them here from overseas or from the factories that made them. All I can ask is, What were we thinking? Oil can only be used once.

What were we thinking?

And I realize that now there is only one thing we can change that will give our descendants any chance of salvaging the best parts of mankind's accumulated experience, art and wisdom—our minds.

MICHAEL C. RUPPERT
November 19, 2009

Michael C. Ruppert is uniquely qualified to write about energy issues, politics, and economics. Over the last ten years he has established an unprecedented track record of accurate energy-related economic, military, and geopolitical prediction, both in his former online newsletter *From The Wilderness* (*FTW*) and in his first book *Crossing the Rubicon: The Decline of the American Empire at the End of the Age of Oil*, (New Society, 2004).

His web site, which ceased operations in late 2006, archived more original reporting on Peak Oil and energy-related topics than any other publication at the time. *FTW* subscribers included major decision makers and members of Congress and the mainstream media. Mike has given more than 200 radio and TV interviews; lectured at more than 20 universities in eight countries; and has addressed California's prestigious Commonwealth Club.

Crossing the Rubicon is in the Harvard Business School and many other university libraries. It is also a book about the attacks of September 11. Although the U.S. government has spent millions of dollars refuting 9/11 "conspiracy theories"—a designation which Ruppert himself disavows—it has never even acknowledged *Rubicon*, with its 1,000 undisputed footnotes. The startling accuracy of the predictions he made in that book and in *FTW* has prompted a recent surge in its popularity among the people who keep America running.

Mike was born in Washington, D.C. into a family with deep connections to the U.S. intelligence community. Prior to graduation from UCLA in 1973 with honors in Political Science, he interned for L.A.'s Chief of Police, during which time he was groomed by officers with connections to the CIA. He later graduated from the LAPD Academy as Valedictorian and was assigned to South Central L.A., where he earned near-perfect evaluations and numerous commendations for investigation, bravery, and attention to duty.

Mike resigned from the LAPD in late 1978 in the wake of death threats he had received after refusing to participate in CIA drug-smuggling operations in the U.S. and went into what he has come to call his "wilderness" for nearly two decades before subsequent revelations and government documents ultimately confirmed his allegations in 1996.

He came into the public spotlight that same year after a confrontation with then-CIA Director John Deutch. His fearless performance on live TV earned him credit for costing Deutch what was considered to be a guaranteed appointment as Secretary of Defense. And despite rare acknowledgement by the major media, Mike's prowess and reputation for investigative journalism is legendary. He broke major investigations, including several stories that later went mainstream.

In May of 2006, for example, after being contacted by Mary Tillman, mother of pro-football star-turned-Army Ranger Pat Tillman, Mike and his Military Affairs editor, Stan Goff (U.S. Army Special Forces, retired) spent days decoding and analyzing 2,000 pages of Army records. The AP, *Washington Post*, and *L.A. Times* (among others) stories which broke the Tillman case open in August and September were virtual cut-and-pastes of *FTW*'s original copyrighted series from June of 2006.

Every congressional handout that forced eventual hearings originated with *FTW*. Nine general officers were disciplined, and many believe that the sudden resignation of Secretary of Defense Donald Rumsfeld was a result of the Tillman scandal. Mike says that perhaps his proudest possession is an autographed copy of *Boots on the Ground by Dusk* by Pat Tillman's mother Mary, thanking him for his help.

Mike's current belief is that attention must now be focused solely on an energy crisis that threatens all of human civilization.

There is no one better equipped to present a comprehensive, fearless, and reliable energy platform that will help America prepare for its greatest challenge ever.

ENDNOTES

Chapter 1

1. http://www.hubbertpeak.com/Hubbert/; "Energy From Fossil Fuels," Marion King Hubbert; *American Association for the Advancement of Science*, Volume 109, No. 2823, February 4, 1949. http://www.hubbertpeak.com/Hubbert/science1949/

2. *Peaking Of World Oil Production: Impacts, Mitigation, & Risk Management*. Robert L. Hirsch, SAIC, Project Leader; Roger Bezdek, MISI; Robert Wendling, MISI: February 2005.

3. *Out Of Gas: The End of the Age of Oil*; David Goodstein (Norton, 2004). Goodstein reaffirmed this position in an October 18, 2004 interview with the *Los Angeles Business Journal* interview entitled "Oil Barren": "The worst case is that Hubbert's Peak occurs and we have a crisis that involves runaway inflation. Not only will gasoline cost more but so will all the commodities made out of petrochemicals. Inflation could even bring worldwide depression. If the economic hit is hard enough we may not be able to rebuild the infrastructure to use something else in place of oil."

Chapter 2

1. The oil supply tsunami alert. Kjell Aleklett, professor in physics. Uppsala University. Uppsala Hydrocarbon Depletion Study Group. President of ASPO, the Association for the Study of Peak Oil & Gas, http://64.233.167.104/search?q=cache:mk9IrfZHQlIJ:www.peakoil.net/Oil_tsunami.html+%22aspo%22+%22in+decline%22+%22oil%22+%22countries%22&hl=en&ct=clnk&cd=2&gl=us&ie=UTF-8

2. Tim Appenzeller, "The End of Cheap Oil"; *National Geographic*, June 2004.

3. http://darwin.bio.uci.edu/~sustain/global/sensem/Forrest98.htm

Chapter 3

1. WorldWatch Institute, *State of the World 2005: Redefining Global Security*. New York: Norton, 107.

2. Data Courtesy Colin Campbell, Ph.D., Association for the Study of Peak Oil, provided to the author, July 2008. Campbell's Excel spreadsheet actually included data on 71 countries. For practical purposes only the largest 20 oil producers have any relevance for discussion.

3. "Kuwait's Burgan Oil Field, World's 2nd Largest, Is 'Exhausted'"; by James Cordahi and Andy Critchlow; *Bloomberg*, Nov. 11, 2005.

4. PEMEX, the Mexican National Oil company; multiple sources including http://www.inteldaily.com/?c=154&a=5140

5. "Mexico's Cantarell oil output falls again in July"; Reuters, Aug. 26, 2008

6. U.S. Department of Energy, Energy Information Administration

7. Hoyos, Carola and Blas, Javier; "World will struggle to meet oil demand"; *The Financial Times*; October 28, 2008.

8. Macalister, Terry;"Key oil figures were distorted by U.S. pressure, says whistleblower; *The Guardian*, Nov. 9, 2009. http://www.guardian.co.uk/environment/2009/nov/09/peak-oil-international-energy-agency

9. Petrobras Discovers World's Third-Largest Oil Field; *Bloomberg*, April 14, 2008.

10. Oil prices enter "super-spike" phase; Reuters, Dec. 13, 2005

11. "The Inevitable Peaking of World Oil Production"; Robert L. Hirsch, *The Atlantic Council of the United States*, Vol. XXV, No 3, October 2005.

12. 2001 author interview with oil geologist and executive Colin Campbell; Paris, 2001.

13. "Fearful EU aims to take energy policy from governments"; by Anthony Brown, *The Times of London*, March 9, 2006: http://www.timesonline.co.uk/article/0,,13509-2076775,00.html. I predicted this in a lecture in Amsterdam in 2003. I reported on it in *From The Wilderness* as it happened; http://www.fromthewilderness.com/free/ww3/031506_peak_prediction.shtml.

14. "High fuel costs spell winter tragedy" by Adam Porter; *Al Jazeera*, Dec. 8, 2004. Adam is a brilliant energy reporter and activist who is quite British and well familiar with the topic. Also: "What they don't want you to know about the coming oil crisis"; by Jeremy Legget, *The Independent*, January 20, 2006: http://news.independent.co.uk/environment/article339928.ece

15. Ibid, Legget.

Chapter 4

1. Interview with *The Energy Bulletin*, May 2, 2008; http://www.energybulletin.net/node/43604

2. "Shell Cuts Oil Reserves Again," *MS-NBC*, March 18, 2004; "Shell Reassesses Reserves Third Time," *Times of London*, April 19, 2004.

3. Ruppert, Michael C.; *Crossing the Rubicon: The Decline of the American Empire at the End of the Age of Oil*; New Society (2004, p.35).

Chapter 5

1. Thanks to http://www.grinzo.com/energy for reporting on this at: http://www.grinzo.com/energy/index.php/2008/04/21/iea-and-lou-on-oil-and-environmental-issues/.

2. "Saudi Aramco to increase Arabian Gulf rigs," *Houston Chronicle*, May 2, 2006.

3. Marianne Lavelle, "The Big Chill: A winter fuel crisis of high prices and shortages could darken homes and factories"; *U.S. News and World Report*, December 19, 2005.

4. Michael C. Ruppert, "The End of the Grid," *From The Wilderness*, January 4, 2006. http://www.fromthewilderness.com/free/ww3/010306_end_grid.shtml.

5. Michael C. Ruppert, "Behind the Blackout," *From The Wilderness*, Aug. 21, 2003. http://www.fromthewilderness.com/free/ww3/082103_blackout.html.

6. U.S. infrastructure needs rise as floods recede; Reuters, July 31, 2008.

7. Jonathan Stemple, "Roads, airports on the block as budgets tighten," Reuters, August 1, 2008.

8. Cauchon, Dennis; *USA Today*; Privately run infrastructure deals dry up"; October 27, 2009. http://www.usatoday.com/news/nation/2009-10-27-Private-infrastructure_N.htm

9. Bureau of Transportation Statistics; U.S. Dept. of Transportation; http://www.bts.gov/publications/white_house_economic_statistics_briefing_room/august_2007/html/highway_vehicle_miles_traveled.html.

10. Michael C. Ruppert, "GlobalCorp," *From The Wilderness*, March 10, 2005; http://www.fromthewilderness.com/free/ww3/031005_globalcorp.shtml.

Chapter 6

1. Bradley Graham, "Commanders Plan Eventual Consolidation of U.S. Bases in Iraq," *Washington Post*, May 22, 2005.

2. Charles Hanley, "U.S. Building Massive Embassy in Baghdad," AP, April 14, 2006.

3. Peter W. Galbraith, "Make Walls, Not War," *New York Times*, October 23, 2007.

4. Original story with map posted at http://live.armedforcesjournal.com/2006/06/1833899 . Map archived at the excellent web site of Mark Robinowitz; http://www.oilempire.us/new-map.html.

5. Leslie H. Gelb, "The Three-State Solution," *New York Times*, November 25, 2003.

6. Doug Smith and Said Rifai, "Foreign Companies Bid to Boost Iraq's Oil Production," *Los Angeles Times*, July 1, 2008.

7. Letter from Senators Schumer and Kerry to Secretary of State Condoleezza Rice, June 23, 2008.

Chapter 7

1. Tanya C. Hsu, "The United States Must Not Neglect Saudi Arabian Investment"; *Saudi-American Forum*, SAF Essay #22, September 23, 2003.
2. Michael C. Ruppert, *Crossing the Rubicon: The Decline of the American Empire at the End of the Age of Oil*, New Society (2004); p.144-145.
3. Energy Information Administration, U.S. Department of Energy; May 2008 Import Highlights: July 28, 2008

Chapter 8

1. *New York Times*, Jan. 11, 2006.
2. Daniel Workman, Suite101.com; "Most Valuable U.S. Food Export is Corn," July 26, 2008; http://import-export.suite101.com/article.cfm/most_valuable_us_food_export_is_corn
3. Joshua Boak and Mike Hughlett, "Corn bonanza won't cut food prices": *Chicago Tribune*, August 13, 2008.
4. Availability of agricultural land for crop and livestock production, Buringh, P. Food and Natural Resources, Pimentel. D. and Hall. C.W. (eds), Academic Press, 1989.
5. Human appropriation of the products of photosynthesis, Vitousek, P.M. et al. Bioscience 36, 1986. http://www.science.duq.edu/esm/unit2-3
6. Land, Energy and Water: the constraints governing Ideal U.S. Population Size, Pimental, David and Pimentel, Marcia. *Focus*, Spring 1991. NPG Forum, 1990. http://www.dieoff.com/page136.htm
7. Constraints on the Expansion of Global Food Supply, Kindell, Henry H. and Pimentel, David. Ambio Vol. 23 No. 3, May 1994. The Royal Swedish Academy of Sciences. http://www.dieoff.com/page36htm
8. The Tightening Conflict: Population, Energy Use, and the Ecology of Agriculture, Giampietro, Mario and Pimentel, David, 1994. http://www.dieoff.com/page69.htm
9. Op. Cit. See note 7.
10. Food, Land, Population and the U.S. Economy, Pimentel, David and Giampietro, Mario. Carrying Capacity Network, 11/21/1994. http://www.dieoff.com/page55.htm
11. Comparison of energy inputs for inorganic fertilizer and manure based corn production, McLaughlin, N.B., et al. Canadian Agricultural Engineering, Vol. 42, No. 1, 2000.
12. Ibid.
13. U.S. Fertilizer Use Statistics. http://www.tfi.org/Statistics/USfertuse2.asp
14. Food, Land, Population and the U.S. Economy, Executive Summary,

Pimentel, David and Giampietro, Mario. Carrying Capacity Network, 11/21/1994. http://www.dieoff.com/page40.htm

15. Ibid.
16. Op. Cit. See note 6.
17. Op. Cit. See note 10.
18. Ibid.
19. Op. Cit. See note 8.
20. Ibid.
21. Ibid.
22. Ibid.
23. Ibid.
24. Op. Cit. See note 14.
25. Ibid.
26. Ibid.
27. Ibid.
28. Op Cit. See note 6.
29. Op Cit. See note 14.
30. Ibid.
31. Ibid.
32. Ibid.
33. Op. Cit. See note 6.
34. Op. Cit. See note 8.
35. Op. Cit. See note 6.
36. Op. Cit. See note 14.
37. Food Consumption and Access, Lynn Brantley, et al. Capital Area Food Bank, 6/1/2001. http://www.clagettfarm.org/purchasing.html
38. Op. Cit. See note 14.
39. Ibid.
40. Op. Cit. See note 8.
41. Ibid.
42. Ibid.
43. Op. Cit. See note 14.
44. Op. Cit. See note 7.
45. Op. Cit. See note 14.
46. Poverty 2002. The U.S. Census Bureau. http://www.census.gov/hhes/poverty/poverty02/pov02hi.html
47. Op. Cit. See note 6.
48. Ibid.
49. *Diet for a Small Planet*, Lappé, Frances Moore. Ballantine Books, 1971-revised 1991. http://www.dietforasmallplanet.com/
50. Op. Cit. See note 8.

51. Ibid.

52. U.S. and World Population Clocks. U.S. Census Bureau. http://www.census.gov/main/www/popclock.html

53. *A Distant Mirror*, Tuckman Barbara. Ballantine Books, 1978.

54. Op. Cit. See note 52.

55. Reuters: Oct 3, 2008—"US Farmers Endure Credit Crunch, worry about 2009."

Chapter 9

1. http://www.rrc.state.tx.us/divisions/og/statistics/production/ogismcon.pdf

2. http://www.rrc.state.tx.us/divisions/og/statistics/drilling/2006/ogdc06an.pdf

3. http://www.rrc.state.tx.us/divisions/og/statistics/drilling/txdrillingstat.pdf

4. *BioScience*, Vol. 44, No. 8, September 1994.

5. Sasha Lilley, "The Dirty Truth About Green Fuel", *CorpWatch*; Posted on June 7, 2006. Printed on June 12, 2006, htttp://www.alternet.org/story/37217/.

6. Clifford Kraus, "Drilling Boom Revives Hopes for Natural Gas," *New York Times*, August 25, 2008.

7. Ibid.

8. (3a)Triple Pundit, "Shale Gas: Energy Boon or Environmental Bane," August 11, 2008, http://www.triplepundit.com/pages/shale-gas-energy-boon-or-envir-003396.php.

9. Triple Pundit, "Boom & Bust, Boon or Bane: Shale Gas Fever Spreads," August 22, 2008, http://www.triplepundit.com/pages/boom-bust-boon-or-bane-shale-g-003434.php.

Chapter 10

1. Wikipedia, Wiki Answers; http://66.102.9.104/search?q=cache:yLepMqqjnz4J:wiki.answers.com/Q/What_are_the_advantages_and_disadvantages_of_ocean_energy+%22tides%22+energy+%22cost%22+%22problems%22&hl=en&ct=clnk&cd=4&gl=us&ie=UTF-8

2. Ocean Energy Solutions; August 20, 2008, http://66.10 2.9.104/search?q=cache:NdFO2SAVPC4J:www.oceanenergysolutions.com/+%22tides%22+energy+%22problems%22&hl=en&ct=clnk&cd=8&gl=us&ie=UTF-8

3. U.S. Department of Energy, "Tracking New Coal-Fired Power Plants," National Energy Technology Laboratory, June 30, 2008—http://www.netl.doe.gov/coal/refshelf/ncp.pdf.

4. Wikipedia, http://www.wikipedia.com, "Carbon Capture and Storage."

5. "Coal's other mess: The solid waste" by Jonathan Thompson; *High Country News*, January 2, 2008: http://www.summitdaily.com/article/20080102/NEWS/225631911.

6. "Coal Fired Plants Take Hits" by K.S. Parthasarathy, *The Tribune of India*, March 14, 2008; http://www.tribuneindia.com/2008/20080314/science.htm#1.

7. Wikipedia: "Clean Coal"; also http://en.wikipedia.org/wiki/S._David_Freeman; as recorded from http://www.grist.org/news/muck/2004/12/03/little-coal/

8. Richard Heinberg, "Peak Coal: Sooner than you think," *The Energy Bulletin*, May 21, 2007; http://www.energybulletin.net/node/29919

9. Rebecca Smith and Daniel Machalaba, "As Utilities Seek More Coal Railroads Struggle to Deliver," *Wall Street Journal*, March 15, 2006, p A1.

10. Dr. Werner Zittel and Jörg Schindler, "Peak coal by 2025 say researchers," *The Energy Bulletin*, March 28, 2007; http://www.energybulletin.net/node/28287

11. U.S. Energy Information Administration; http://www.eia.doe.gov; http://www.eia.doe.gov/emeu/recs/recs2001/enduse2001/enduse2001.html

12. Mark Robinowitz, "Peak Coal and Mountaintop Removal," http://www.oilempire.us/peak-coal.html.

13. Steven Mufson, "Democrats Push Coal-to-Liquids Energy Plan, *The Washington Post*, Wednesday June 13, 2007.

14. "A Dirty Little Secret Canada's Global Warming Engine"; http://www.tarsandswatch.org; and the U.S. Department of Energy; http://www.doe.gov.

15. "Tar Sands: The Most Destructive project on Earth"; http://www.treehugger.com/files/2008/02/tar-sands-most-destructive-project.php

16. Wikipedia, Tar Sands

17. Ibid.

18. "Alternative Energy Sources," by Walter Youngquist, Consulting Geologist, Eugene, Oregon, October 2000; http://www.hubbertpeak.com/Youngquist/altenergy.htm.

19. Ken Salazar, "Heedless Rush to Oil Shale"; *The Washington Post*, Tuesday July 15, 2008, P A19.

20. "WMC Ideally Placed to Deal with Increased Uranium Demand." *AZo Journal of Materials Online* (2004-12-16). (As reported on Wikipedia).

Chapter 11

1. Larry Rohter, "Shipping Costs Start to Crimp Globalization," *New York Times*, Aug 3, 2008.
2. Associated Press, July 24, 2008.
3. *Los Angeles Times*, June 30, 2008
4. Reuters, July 24 2008.
5. *The Daily Times*, June 16, 2006
6. *New York Times*, June 21, 2008
7. I strongly recommend the amazing documentary film *The Power of Community* from http://www.communitysolution.org.
8. Dale Allen Pfeiffer, "Drawing Lessons From Experience," Parts I & II, *From The Wilderness*, http://www.fromthewilderness.com/free/ww3/111703_korea_cuba_1.html http://www.fromthewilderness.com/free/ww3/120103_korea_2.html

Chapter 12

1. http://land.netonecom.net/tlp/ref/federal_reserve.shtml
2. http://www.agric.nsw.gov.au/reader/tg_Size_and_Weight.htm
3. "Cars in China," *The Economist*, June 2, 2005; http://www.economist.com/world/displaystory.cfm?story_id=4032842
4. http://china-economics-blog.blogspot.com/2008/07/cars-in-china-more-research.html
5. Michael C. Ruppert, "Paris Peak Oil Conference Reveals Deepening Crisis," *From The Wilderness*, May 30, 2003; http://www.fromthewilderness.com/free/ww3/053103_aspo.html.
6. "Fast German Energy Facts," http://www.solarbuzz.com; http://www.solarbuzz.com/fastfactsgermany.htm
7. Carol Gulyas, "Financing Renewable Energy: Feed-in Tariff Introduced in Congress"; *Clean Technica*, http://cleantechnica.com, July 6, 2008.
8. Ashley Seager, "Germany Sets Shining Example in Providing a Harvest for the World"; *The Guardian*, July 23, 2007.
9. Jenna Orkin, "Wagging the Dog: Economic Growth Leaves Water, People and Food Supplies in the Dust," *From The Wilderness*, October 6, 2006; http://www.fromthewilderness.com/members/101606_leaves_water.shtml

Chapter 13

1. James Kirkup, "U.S., U.K. Waged War on Iraq Because of Oil, Blair Advisor Says"; *Bloomberg News*, May 1, 2003. http://quote.bloomberg.com/apps/news?pid=10000087&sid=ahJS35XsmXGg&refer=top

2. Peter Baker, "Bush Say US Pullout Would Let Iraq Radicals Use Oil as Weapon; *The Washington Post*, Nov. 5. 2006

3. Peter Beaumont and Joanna Walters, "Greenspan admits Iraq was about oil, as deaths put at 1.2m"; *The Guardian*, Sept. 16, 2007.

4. Glenn Kessler, "US Decision on Iraq Has Puzzling Past"; *The Washington Post*, Jan. 12, 2003.

5. I documented this entire history in *Crossing the Rubicon* from a multitude of major press sources and official announcements. Confirmation of his relationship with Unocal was obtained through a personal profile downloaded from the U.S. State Department web site in late 2001.

6. National Security Strategy of the United States, http://www.whitehouse.gov/nsc/printnssall.html

7. Michael C. Ruppert, *Crossing the Rubicon: the Decline of the American Empire at the End of the Age of Oil*; New Society (2004), p.473

8. Ibid

9. Op cit; Ruppert, p. 575

10. http://www.oildepletionprotocol.org/theprotocol

11. Richard Heinberg, *The Oil Depletion Protocol*, New Society Publishers, 2006; p. 79

Chapter 14

1. Lercah, Daniel, *Post Carbon Cities: Planning for Energy and Climate Uncertainty*, Post Carbon Institute, 2007, p. 4.

2. U.S. Department of Energy, Strategic Petroleum Reserve; http://fossil.energy.gov/programs/reserves

3. The United States Federal Reserve Bank; National Economic Trends, "China's Strategic Petroleum Reserve: A Drop in the Bucket," January 2007.

4. Wikipedia.com: "global strategic petroleum reserves"

5. http://www.fossil.energy.gov/programs/reserves/spr/expansion-eis.html

6. U.S. Department of Energy, Strategic Petroleum Reserve; http://fossil.energy.gov/programs/reserves

7. Dale Allen Pfeiffer, "Oil Shortages Look Certain by 2007," Feb. 19, 2004; From The Wilderness Publications; http://www.fromthewilderness.com/free/ww3/022304_lng_shortages.html

8. http://www.fromthewilderness.com/free/ww3/111707_oil_depletion.shtml

9. Michael C. Ruppert, "Economic Alert #4—The Abyss Awaits," From The Wilderness Publications, Inc.; June 14, 2006; http://www.fromthewilderness.com/free/ww3/061406_abyss_awaits.shtml

Chapter 15

1. Shachtman, Noah; *WIRED* Magazine; November 11, 2009; "Afghanistan's Oil Binge: 22 Gallons of Fuel Per Soldier Per Day. http://www.wired.com/dangerroom/2009/11/afghanistans-oil-binge-22-gallons-of-fuel-per-soldier-per-day/

Epilogue

1. O'Grady, Sean; "Surge in the price of oil is a fresh threat to recovery"; *The Independent*, November 13th 2009; http://www.independent.co.uk/news/business/news/surge-in-the-price-of-oil-is-a-fresh-threat-to-recovery-1819879.html

2. Quigley, Christopher M.; Re-evolution and the steady state of M. King Hubbert, June 12, 2007. Financial Sense University; http://www.financialsense.com/fsu/editorials/quigley/2007/0612.html

3. Hickerson, Robert L.; "Hubbert's Prescription for Survival, A Steady State Economy, March 1, 1995; http://www.hubbertpeak.com/hubbert/hubecon.htm

4. Ibid

INDEX